T0341182

Design and Analysis of Experiments and Observational and Studies using R

CHAPMAN & HALL/CRC
Texts in Statistical Science Series

Joseph K. Blitzstein, *Harvard University, USA*
Julian J. Faraway, *University of Bath, UK*
Martin Tanner, *Northwestern University, USA*
Jim Zidek, *University of British Columbia, Canada*

Recently Published Titles

Bayesian Networks
With Examples in R, Second Edition
Marco Scutari and Jean-Baptiste Denis

Time Series
Modeling, Computation, and Inference, Second Edition
Raquel Prado, Marco A. R. Ferreira and Mike West

A First Course in Linear Model Theory, Second Edition
Nalini Ravishanker, Zhiyi Chi, Dipak K. Dey

Foundations of Statistics for Data Scientists
With R and Python
Alan Agresti and Maria Kateri

Fundamentals of Causal Inference
With R
Babette A. Brumback

Sampling
Design and Analysis, Third Edition
Sharon L. Lohr

Theory of Statistical Inference
Anthony Almudevar

Probability, Statistics, and Data
A Fresh Approach Using R
Darrin Speegle and Brain Claire

Bayesian Modeling and Computation in Python
Osvaldo A. Martin, Raviv Kumar and Junpeng Lao

Bayes Rules!
An Introduction to Applied Bayesian Modeling
Alicia Johnson, Miles Ott and Mine Dogucu

Stochastic Processes with R
An Introduction
Olga Korosteleva

Introduction to Design and Analysis of Scientific Studies
Nathan Taback

Practical Time Series Analysis for Data Science
Wayne A. Woodward, Bivin Philip Sadler and Stephen Robertson

For more information about this series, please visit:
https://www.routledge.com/Chapman--HallCRC-Texts-in-Statistical-Science/book-series/
CHTEXSTASCI

Design and Analysis of Experiments and Observational and Studies using R

Nathan Taback

CRC Press
Taylor & Francis Group
Boca Raton London New York

CRC Press is an imprint of the
Taylor & Francis Group, an **informa** business

A CHAPMAN & HALL BOOK

First edition published 2022
by CRC Press
6000 Broken Sound Parkway NW, Suite 300, Boca Raton, FL 33487-2742

and by CRC Press
4 Park Square, Milton Park, Abingdon, Oxon, OX14 4RN

CRC Press is an imprint of Taylor & Francis Group, LLC

Library of Congress Cataloging-in-Publication Data

Names: Taback, Nathan, author.
Title: Design and Analysis of Experiments and Observational and Studies using R / Nathan Taback.
Description: First edition. | Boca Raton : CRC Press, 2022. | Series:
Chapman & Hall/CRC texts in statistical science | Includes
bibliographical references and index.
Identifiers: LCCN 2021048076 (print) | LCCN 2021048077 (ebook) | ISBN
9780367456856 (hardback) | ISBN 9781032219844 (paperback) | ISBN
9781003033691 (ebook)
Subjects: LCSH: Experimental design. | Research--Statistical methods.
Classification: LCC QA279 .T283 2022 (print) | LCC QA279 (ebook) | DDC
519.5/7--dc23/eng/20211130
LC record available at https://lccn.loc.gov/2021048076
LC ebook record available at https://lccn.loc.gov/2021048077

ISBN: 978-0-367-45685-6 (hbk)
ISBN: 978-1-032-21984-4 (pbk)
ISBN: 978-1-003-03369-1 (ebk)

DOI: 10.1201/9781003033691

Typeset in LM Roman
by KnowledgeWorks Global Ltd.

Publisher's note: This book has been prepared from camera-ready copy provided by the authors.

To Monika, Adam, and Oliver.

Contents

List of Tables

List of Figures

Symbols and Abbreviations

Symbol Description

$N(\mu, \sigma^2)$ Normal distribution with mean μ and variance σ^2.

p.d.f.   Probability density function.

CDF   Cumulative distribution function.

z_p p^{th} quantile of the $N(0,1)$ distribution.

$\Phi(\cdot)$ The CDF of the $N(0,1)$.

$\phi(\cdot)$ The density function of the $N(0,1)$.

χ_n^2 Chi-squared distribution on n degrees of freedom.

t_n Student's t distribution on n degrees of freedom.

$F_{m,n}$ F distribution with numerator degrees of freedom m and denominator degrees of freedom n.

i.i.d.   Independent and identically distributed.

$E(X)$ Expected value of X.

$Var(X)$ Variance of X.

s.e.$(\hat{\beta})$ Standard error of $\hat{\beta}$

$(m_1, m_2)'$ Transpose of (m_1, m_2)

S_x Sample standard deviation of $x = (x_1, \ldots, x_n)$

FWER   Family wise error rate

$F_{a,b}(\delta)$ Non-central F distribution with numerator and denominator degrees of freedom a, b and non-centrality parameter δ

Preface

This book grew out of my course notes for a twelve-week course (one term) on the Design of Scientific Studies at the University of Toronto. A course for senior undergraduates and applied Masters students who have completed courses in probability, mathematical statistics, and regression analysis.

I started writing my own notes for two reasons: (1) to expose students to the foundations of classical experimental design and design of observational studies through the framework of causality. Causal inference provides a framework for investigators to readily evaluate study limitations and draw appropriate conclusions; (2) because I believe statistics students should have increased exposure to building and using computational tools, such as simulation, and real data in the context of these topics, and R is a fantastic language to use for this purpose. The book uses R to implement designs and analyse data. It's assumed that the reader has taken basic courses in probability, mathematical statistics, and linear models, although the essentials are reviewed briefly in the first chapter. Some experience using R is helpful although not essential. I assume that readers are familiar with standard base R and **tidyverse** syntax. In the course at the University of Toronto, students are given learning resources at the beginning of the course to review these R basics, although most students have had some exposure to computing with R.

Nathan Taback

Toronto, Ontario, Canada

September, 2021

Organization of the book

The structure of each chapter presents concepts or methods followed by a section that shows readers how to implement these in R. These sections are labeled "*Computational Lab: Topic*", where "*Topic*" is the topic that is implemented in R.

Software information and conventions

One of the unique features of this book is the emphasis on simulation and computation using R. R is wonderful because of the many open source packages available, but this can also lead to confusion about which packages to use for a task. I have tried to minimize the number of packages used in the book. The set of packages loaded on startup by default is

```
getOption("defaultPackages")
```

```
## [1] "datasets"  "utils"      "grDevices" "graphics"
## [5] "stats"     "methods"
```

plus base. If a function from a non-default library is used, then this is indicated by pkg::name instead of

```
library(pkg)
name
```

This should make it clear which package a user needs to load before using a function.

Information on the R version used to write this book is below.

```
version
```

```
##                  -
## platform         x86_64-apple-darwin17.0
## arch             x86_64
## os               darwin17.0
## system           x86_64, darwin17.0
## status
## major            4
## minor            1.1
## year             2021
## month            08
## day              10
## svn rev          80725
```

```
## language       R
## version.string R version 4.1.1 (2021-08-10)
## nickname       Kick Things
```

The packages used in writing this book are:

```
library(tidyverse)
library(knitr)
library(kableExtra)
library(reticulate)
library(janitor)
library(latex2exp)
library(gridExtra)
library(broom)
library(patchwork)
library(crosstable)
library(agridat)
library(FrF2)
library(pwr)
library(emmeans)
library(DiagrammeR)
library(abind)
library(magic)
library(BsMD)
library(scidesignR)
```

R code

Whenever possible the R code developed in this book is written as a function instead of a series of statements. "Functions allow you to automate common tasks in a more powerful and general way than copy-and-pasting" [Wickham and Grolemund, 2016]. In fact, I have taken the approach that whenever I've copied and pasted a block of code more than twice then it's time to write a function.

The value an R function returns is the last value evaluated. return() can be used to return a value before the last value. Many of the functions in this book use return() to make code easier to read even when the last value of the function is returned.

R 4.1.0 now provides a simple native forward pipe syntax |>. The simple form of the forward pipe inserts the left-hand side as the first argument in the right-hand side call. The pipe syntax used in this book is %>% from the magrittr library. Most of the code in this book should work with the native pipe |>, although this has not been thoroughly tested.

Data sets

The data sets used in this book are available in the R package scidesignR, and can be installed by running

```
install.packages("devtools")
devtools::install_github("scidesign/scidesignR")
```

Acknowledgments

I would like to thank all the students and instructors who used my notes and provided valuable feedback. Michael Moon provided excellent assistance with helping me develop many of the exercises. Monika, Adam, and Oliver, as usual, provided sustained support throughout this project.

1

Introduction

The amount of data collected across domains in science, medicine, business, humanities, engineering, and the arts in the past decade has exploded. Sensors and internet-enabled devices, such as smart phones, deliver data to databases stored in the cloud; scientific experiments at The Large Hadron Collider (LHC) particle accelerator at CERN produce approximately 30 petabytes (30 million gigabytes) of data annually [Suthakar et al., 2016], Facebook processes 2.5 billion pieces of content and more than 500 TB of data each day [Techcrunch, 2012], and health information is often stored in electronic health records. Data collection has been an important part of science throughout history, but these examples show that people working in scientific, business, and other areas will encounter data collection, so understanding it is important in this data-driven era.

A *study* is an empirical investigation of the relationship between *effects* and *treatments* . This book introduces the reader to the fundamental principles of study design using computation and simulation. In general, a *study design* includes all activities that precede data analysis including what data to collect, and where, when, and how it will be collected. An important aspect of study design is framing a scientific question, testable with data, to address an issue. Addressing the issue can often lead to knowledge helpful in understanding or improving a process or product. The interplay between how data should be collected, and the strength of the answers to the questions (i.e., conclusions) will be covered throughout this book.

1.1 Planning and Implementing a Study

In this section some guidelines and definitions for planning and implementing a study are given.

State study objectives

What do the stakeholders hope to learn from the experiment? Evaluating the impact of web page design on sales is an example of a **study objective**. A study may have multiple objectives.

DOI: 10.1201/9781003033691-1

Select responses

Which variable(s) will be used to evaluate a study's objective? In the previous example sales or number of items sold could be used as a **response variable** to evaluate the objective.

Select study treatments and other covariates

A **factor** is a variable thought to affect the response variable. Factors can be **quantitative** or **qualitative**. Quantitative factors, such as temperature and time, take values over a continuous range. Qualitative factors, such as sex or marital status, take on a discrete number of values. The values of a factor are called **levels**. A **treatment** is a level of a factor in the case of one factor, and a combination of factor levels in the case of more than one factor.

Develop a study plan

Use the principles discussed in this book to develop a plan of the how the data will be collected.

Data collection

A plan describing the actual values of the factors used and how the response will be collected is very important.

Data analysis

A data analysis to evaluate the relationship between the factors and response variables appropriate for the experimental plan should be carried out.

Make conclusions/recommendations

Data analysis can be used to draw conclusions and make recommendations based on the effect of treatments on response variables.

1.2   Questions

Data can be used to answer questions such as:

- Does a new treatment, compared to no treatment, lower the death rate in people that have a ceratin disease?
- Does modifying the web page of an e-commerce site increase sales?
- Does increasing the amount of a chemical additive increase the stability of a product?
- Does changing body posture increase success in life?

1.3 Types of Study Investigations

Investigators can be defined as a group of individuals who carry out an investigation. Scientists, business analysts, or engineers often conduct empirical investigations to answer questions they hope will lead to new knowledge. This book is about designing two types of empirical investigations: **experiments** and **observational studies**. An **experiment** is an investigation where all the activities preceding data analysis, such as selection of experimental units and assignment of treatments to these units, can be controlled by the investigators. In an **observational study** some of these activities cannot be controlled by the investigators, especially the allocation of experimental units to treatments.

Example 1.1. The director of analytics for a political candidate wants visitors to the candidate's website to register with the website by clicking a button on the website. The campaign staff debates what text the button should display and which images should be displayed on the landing web page. The campaign team is considering three buttons, and four different images (see Figure 1.1).

FIGURE 1.1: Web Page Buttons and Images

Framing a scientific question: A scientific question articulates a hypothesis that can be evaluated by collecting data. How can this issue be framed as a scientific question? Which button and which image lead to the highest number of visitors signing-up is an example of a scientific question.

Data collection: Empirical investigations involves collecting data. An important decision is what data to collect, and when and how it should be collected. For example, should the **outcome measure** used to evaluate success be number of clicks or number of people that clicked and signed-up? The campaign could change the landing page every few days to evaluate the 12 combinations (3 buttons × 4 images), or visitors could be randomly assigned to one of the 12 landing pages over a week. Which data collection method will provide a *stronger* answer to the question?

1.4   Strength of Answers

In Example 1.1 the two methods discussed to collect data would generate a data set with the same variables, albeit different values of the variables. So, why worry about which method to use since the same statistical analysis can be done on both data sets? The answer lies in the strength of the conclusions that can be drawn from each data set. If the campaign changed the landing page every few days, then it's plausible that the time a particular landing page is available could impact the characteristics of users that potentially view the page and ultimately decide to sign-up. Conversely, if all the pages were available at the same time, then only chance impacts which page a user views. This implies that any differences between the visitors that view landing pages should be random.

Experiments outline a *plan* for data collection and analysis, whereas undesigned or exploratory studies don't outline a plan that can lead to data used to draw *sound* conclusions.

1.5   Why Design Scientific Studies?

Why should scientific studies be designed when so much data is already available and routinely collected? Available data is often insufficient and inappropriate to answer important questions. Dexamethasone is an inexpensive generic drug that some doctors thought could be useful in treating COVID-19 patients, but others disagreed. Studies of dexamethasone used to treat related illnesses suggested it might harm instead of help. A study designed by British investigators demonstrated that dexamethasone was able to reduce deaths, and explained when a patient with COVID-19 should be treated. Dexamethasone has become a staple of treatment for COVID-19. Even though data was already available on using dexamethasone, there was no data available on its use in COVID-19 patients. Designing scientific studies can lead to the collection of high quality data which in turn leads to more accurate information compared to having a large amount of lower quality data [Tufekci, 2021].

Are statistical sampling and randomization still relevant in the era of Big Data? This question was asked by Professor Xiao-Li Meng [Harvard University, Department of Statistics, 2015]. Meng considers the following: If we want to estimate a population mean, which data set would give us more accurate results, a 1% simple random sample or a data set containing self-reports of the quantity of interest that covers 95% of the population?

The total error is captured by the mean square error (MSE). The mean square error of an estimator $\hat{\theta}$ of θ is,

$$MSE = E\left(\hat{\theta} - \theta\right) = Var\left(\hat{\theta}\right) + \left(E\left(\hat{\theta} - \theta\right)\right)^2.$$

In other words, $Total\, Error = Variance + Bias^2$. The term $E\left(\hat{\theta} - \theta\right)$ is called the bias of the estimator $\hat{\theta}$. If the bias is 0 then the estimator is called unbiased.

Suppose we have a finite population of measurements $\{x_1, ..., x_N\}$ of some quantity, say the total number of hours spent on the Internet during a one-year period for every person in Canada. The population of Canada in 2015 is $N = 35,749,600$ or approximately 35.8 million people [Statistics Canada, The Daily, 2015]. In order to estimate the mean number of hours spent on the Internet is it better to take a simple random sample of 100 people and estimate the mean number of hours spent on the Internet or use a large database (much larger than the random sample) that contains self-reports of hours spent on the Internet?

Suppose that $\bar{x}_a$ is the sample mean from the database and $\bar{x}_s$ is the mean from the random sample. Meng [2014] shows that $MSE\left(\bar{x}_a\right) < MSE\left(\bar{x}_s\right)$ if

$$f_a > \frac{n_s \rho_N^2}{1 + n_s \rho^2},$$

where f_a is the fraction of the population in the database, ρ is the correlation between the response being recorded and the recorded value, and n_s is the size of the random sample. For example, if $n_s = 100$ the database would need over 96% of the population if $\rho = 0.5$ to guarantee that $MSE\left(\bar{x}_a\right) < MSE\left(\bar{x}_s\right)$. In our example this would require a database with 34,319,616 Canadians.

This illustrates the main advantage of designing and collecting a probabilistic sample and the danger of putting faith in "Big Data" simply because it's Big!

2

Mathematical Statistics: Simulation and Computation

2.1 Data

Statistics and data science begin with data collection. A primary focus of study design is how to *best* collect data that provides *evidence* to *support* answering *questions*. Questions are in the form, if X changes does Y change?

In some contexts data collection can be designed so that an investigator can control how people (or in general experimental units) are assigned to a ("treatment") group, and in other contexts it's neither feasible nor ethical to assign people to groups that are to be compared. In the latter case the only choice is to use data where people ended up in groups, but we don't know the probability of belonging to a group.

2.1.1 Computation Lab: Data

This section contains a very brief overview of using R to read, manipulate, summarize, visualize, and generate data. The computation labs throughout this book will use a combination of **base** R [R Core Team, 2021] and the **tidyverse** collection of packages [Wickham, 2021] to deal with data.

2.1.1.1 Reading Data into R

When data is collected as part of a study, it will almost always be stored in computer files. In order to use R to analyse data, it must be read into R. The data frame data structure is the most common way of storing data in R [Wickham, 2019].

Example 2.1. `nhefshwdat1.csv` is a text file in comma separated value format. `nhefshwdat1.csv` can be imported into an R data frame using `read.csv()` and stored as an R data frame `nhefsdf`. The `scidesignR` package includes this text file and a function `scidesignR::scidesignR_example()`

that makes the file easy to access within the package. The first six columns
and three rows are shown.

```
nhefshwdat1_filepath <-
  scidesignR::scidesignR_example("nhefshwdat1.csv")
nhefsdf <- read.csv(nhefshwdat1_filepath)
head(nhefsdf[1:6], n = 3)
```

```
##   X seqn age sex race education.code
## 1 1  233  42   0    1              1
## 2 2  235  36   0    0              2
## 3 3  244  56   1    1              2
```

The variable names in the data set can be viewed using `colnames()`.

```
colnames(nhefsdf)
```

```
##  [1] "X"              "seqn"          "age"
##  [4] "sex"            "race"          "education.code"
##  [7] "smokeintensity" "smokeyrs"      "exercise"
## [10] "active"         "wt71"          "qsmk"
## [13] "wt82_71"        "pqsmkobs"      "strat1"
## [16] "strat2"         "strat3"        "strat4"
## [19] "strat5"         "stratvar"
```

2.1.1.2   Manipulating Data in R

Raw data from a study is usually not in a format that is ready to be analysed,
so it must be manipulated or wrangled before analysis. R (often called base R)
has many functions to manipulate data. R's default packages and the `dplyr`
library [Wickham et al., 2021b], which is part of the `tidyverse` collection of
packages, provide many functions to solve common data manipulation problems
[Wickham and Grolemund, 2016]. Both will be used together and throughout
this book.

Example 2.2. The education variable `nhefsdf$education.code` was stored
as an integer but is an ordinal categorical variable with five levels. If we want
to recode this into a variable that has two levels with 4 and 5 mapped to 1,
and 1, 2, 3 mapped to 0 then we can use `dplyr::recode()`.

`nhefsdf$education.code`'s data type is an *integer type*, where the integers represent education categories. Factors are special R objects used to handle categorical variables in R. `as.factor()` is a function that encodes a vector as a factor. `levels()` extracts the levels of the factor.

```
levels(as.factor(nhefsdf$education.code))
```

```
## [1] "1" "2" "3" "4" "5"
```

```
edcode <- dplyr::recode(
  as.factor(nhefsdf$education.code),
  "5" = 1,
  "4" = 1,
  "3" = 0,
  "2" = 0,
  "1" = 0
)
levels(as.factor(edcode))
```

```
## [1] "0" "1"
```

2.1.1.3 Summarizing Data using R

This section will provide a brief overview of summarizing data using R. Wickham and Grolemund [2016] provides an excellent discussion of tools to summarize data using R.

Data analysis involves exploring data quantitatively and visually. Quantitative summaries are usually presented in tables and visual summaries in graphs. The `dplyr` library contains many functions to help summarise data. A common way to express short sequences of multiple operations in tidyverse libraries is to use the pipe `%>%` [Wickham and Grolemund, 2016].

The `ggplot2` library [Wickham et al., 2021a], part of the `tidyverse` collection, can be used to create high quality graphics. `ggplot2` and `dplyr` can both be loaded by loading the `tidyverse` package.

2.1.1.3.1   R pipes

This R code

```
nhefsdf %>%
  dplyr::count(education.code)
```

is equivalent to

```
dplyr::count(nhefsdf, education.code)
```

`dplyr::count()` returns a data frame. So, the resulting data frame can be "piped" (i.e., `%>%`) into another `tidyverse` function.

The code chunk below loads the `tidyverse` library (`library(tidyverse)`) so there is no need to use `pkg::name` notation. Below, the data frame from `nhefsdf %>% count(education.code)` is used as the data frame argument in `mutate(pct = round(n/sum(n)*100,2))`. `dplyr::mutate()` is used to create a new column in the data frame returned by `nhefsdf %>% count(education.code)`, and `dplyr::summarise()` summarises `nhefsdf` by several functions `n()`—the number of observations, `mean(age)`—mean of `age` column, `median(age)`—median of `age` column, and `sd(age)`—standard deviation of `age`.

```
library(tidyverse)

nhefsdf %>%
  count(education.code) %>%
  mutate(pct = round(n/sum(n)*100,2))
```

```
##   education.code   n    pct
## 1              1 291  18.58
## 2              2 340  21.71
## 3              3 637  40.68
## 4              4 121   7.73
## 5              5 177  11.30
```

```
summarise(
  nhefsdf,
  N = n(),
  mean = mean(age),
  median = median(age),
  sd = sd(age)
)
```

```
##       N  mean median     sd
## 1 1566 43.66     43 11.99
```

The `patchwork` library [Pedersen, 2020] can be used to combine separate plots created with `ggplot2` into the same graphic. Wickham et al. [2009] provides an excellent introduction to `ggplot2`.

```
library(patchwork)

barpl <- nhefsdf %>%
  ggplot(aes(education.code)) +
  geom_bar()

histpl <- nhefsdf %>%
  ggplot(aes(age)) +
  geom_histogram(binwidth = 10, colour = "black", fill = "grey")

barpl + histpl
```

2.1.1.4 Generating Data in R

Study design often involves generating data. Generating data sets can involve replicating values to create variables, combining combinations of variables, or sampling values from a known probability distribution.

Example 2.3. A data frame called `gendf` that contains 10 rows and variables `trt` and `value` will be computed. `trt` should have five values equal to `trt1`, five equal to `trt2`, and randomly assigned to the data frame rows; `value` should be a random sample of 10 values from the numbers 1 through 1,000,000. This example will use a random number seed (`set.seed()`) so that it's possible to *reproduce* the same values if the code is run again.

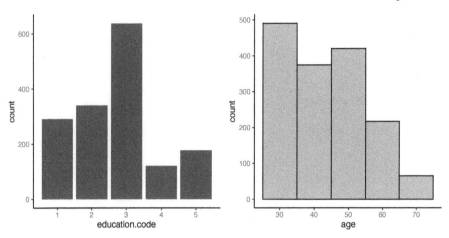

FIGURE 2.1: ggplot2 Examples with patchwork

```
set.seed(11)
gendf <- data.frame(trt = sample(c(rep("trt1", 5), rep("trt2", 5))),
                    value = sample(1:1000000, 10))
head(gendf, n = 3)
```

```
##     trt  value
## 1 trt2 387134
## 2 trt1 900071
## 3 trt2 883541
```

Example 2.4. An investigator is designing a study with three subjects and three treatments, A, B, C. Each subject is required to recieve each treatment, but the order is random.

The code below shows how **expand.grid()** is used to create a data frame from all combinations of **subj** and **treat**. **run_order** is a random permutation of $1, \ldots, 9$ that is sorted using **arrange()**.

```
set.seed(11)

expand.grid(subj = 1:3, treat = c("A", "B", "C")) %>%
  mutate(run_order = sample(1:9)) %>%
  arrange(run_order) %>%
  head(n = 3)
```

```
##   subj treat run_order
## 1    3     A         1
## 2    1     A         2
## 3    2     C         3
```

2.2 Frequency Distributions

A **frequency distribution** of data $x_1, x_2, , \ldots, x_n$ is a function that computes the frequency of each distinct x_i, if the data are discrete, or the frequency of x_i's falling in an interval, if the data are continuous.

2.2.1 Discrete and Continuous Data

How many subjects died and how many survived in a medical study? This is called the *frequency distribution* of survival. An example is shown in Table 2.1. Death is a *discrete random variable*: it can take on a finite number of values (i.e., 0 (alive) and 1 (dead)), and its value is random since the outcome is uncertain and occurs with a certain probability.

Age is an example of a *continuous random variable* since it can take on a continuum of values rather than a finite or countably infinite number as in the discrete case.

TABLE 2.1: Frequency Distribution of Survival

Death	Number of Subjects	Mean Age (years)
0	17	33.35
1	3	36.33

Example 2.5 (COVID-19 Treatment Study). A 2020 National Institutes of Health (NIH) sponsored study to evaluate if a treatment for COVID-19 is effective planned to enroll 20 subjects and *randomly* assign 10 to the treatment and 10 to a fake treatment (*placebo*). Effectiveness was evaluated by the proportion of participants who died from any cause or were hospitalized during the 20-day period from and including the day of the first (confirmed) dose of study treatment [Clinicaltrial.gov, 2020].

Simulated data for this study data is shown in Table 2.2. trt records if a subject received treatment TRT or placebo PLA, die_hosp is 1 if a participant died or was hospitalized in the first 20 days after treatment, and 0 otherwise,

TABLE 2.2: COVID-19 Study Data

patient	trt	die_hosp	age
1	TRT	0	38
2	PLA	0	33
3	PLA	0	30
4	TRT	1	39
5	TRT	0	26
6	TRT	0	37

and `age` is the subject's age in years. The data is stored in the data frame `covid19_trial`.

2.2.2 Computation Lab: Frequency Distributions

2.2.2.1 Discrete Data

The data for Example 2.5 is in the data frame `covid19_trial`. `dplyr::glimpse` can be used to print a data frame.

```
dplyr::glimpse(covid19_trial)
```

```
## Rows: 20
## Columns: 4
## $ patient  <int> 1, 2, 3, 4, 5, 6, 7, 8, 9, 10, 11, 1~
## $ trt      <chr> "TRT", "PLA", "PLA", "TRT", "TRT", "~
## $ die_hosp <fct> 0, 0, 0, 1, 0, 0, 0, 0, 0, 0, 0, 1, ~
## $ age      <dbl> 38, 33, 30, 39, 26, 37, 29, 32, 29, ~
```

`dplyr::group_by()` is used to group the data by `die_hosp` - if a subject is dead or alive, and `dplyr::summarise()` is used to compute the number of subjects, and mean age in each group (see Table 2.1).

```
covid19_trial %>%
  group_by(Death = die_hosp) %>%
  summarise(`Number of Subjects` = n(),
            `Mean Age (years)` = mean(age))
```

```
## # A tibble: 2 x 3
##   Death `Number of Subjects` `Mean Age (years)`
```

```
##    <fct>                 <int>              <dbl>
## 1 0                       17                33.4
## 2 1                        3                36.3
```

The code below was used to create Figure 2.2, which shows a plot of the frequency distribution of survival using `ggplot2::ggplot()` to create a bar chart.

```
covid19_trial %>%
  ggplot(aes(die_hosp)) +
  geom_bar(colour = "black", fill = "grey") +
  scale_y_discrete(limits = as.factor(1:17)) +
  ylab("Number of patients") + xlab("Survival")
```

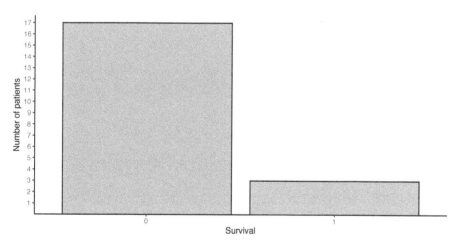

FIGURE 2.2: Distribution of Survival—COVID-19 Study

Is the survival distribution the same in the treatment (**TRT**) and placebo (**PLA**) groups? Table 2.3 shows the frequency distribution and the *relative frequency distribution* of survival in each group. `janitor::tabyl()`, `janitor::adorn_*()` (There are several **adorn** functions for printing— `adorn_*()` refers to the collection of these functions) functions from the janitor [Firke, 2021] library can be used to obtain the cross-tabulation of `die_hosp` and `trt` shown in Table 2.3.

```
library(janitor)
covid19_trial %>% tabyl(die_hosp,trt) %>%
  adorn_totals(where=c("row", "col")) %>%
  adorn_percentages(denominator = "col") %>%
  adorn_ns(position = "front")
```

TABLE 2.3: Survival Distribution—COVID-19 Study

die_hosp	PLA	TRT	Total
0	9 (0.9)	8 (0.8)	17 (0.85)
1	1 (0.1)	2 (0.2)	3 (0.15)
Total	10 (1)	10 (1)	20 (1)

Let n_{ij} be the frequency in the ith row, j^{th} column of Table 2.3. $n_{\bullet1} = \sum_i n_{i1}$, so, for example, $f_{11} = n_{11}/n_{\bullet1}$, and $f_{21} = n_{21}/n_{\bullet1}$ specify the *relative frequency distribution* of survival in the placebo group. If a patient is randomly drawn from the placebo group, there is a 10% chance that the patient will die. This is the reason that the relative frequency distribution is often referred to as an *observed* or *empirical probability distribution*.

The code below was used to create Figure 2.3 using ggplot(). The frequency distributions are shown using a side-by-side bar chart.

```
covid19_trial %>%
  mutate(die_hosp = as.factor(die_hosp)) %>%
  ggplot(aes(die_hosp)) +
  geom_bar(aes(fill=trt), position = "dodge") +
  scale_fill_manual("Treatment",
                    values = c("TRT" = "black",
                               "PLA" = "lightgrey")) +
  scale_y_continuous(limits = c(0,9), breaks = 0:9) +
  ylab("Number of patients") +
  xlab("Survival")
```

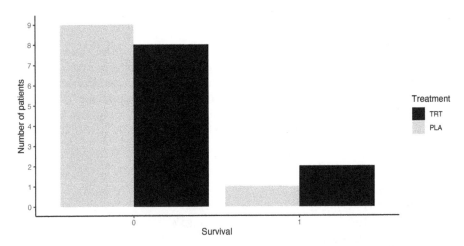

FIGURE 2.3: Distribution of Survival by Group—COVID-19 Study

2.2.2.2 Continuous Data

What is the distribution of patient's age in Example 2.5? Is the distribution the same in each treatment group?

The code below was used to create Table 2.4—a numerical summary of the age distribution for each treatment group.

```
covid19_trial %>% group_by(trt) %>%
  summarise(n = n(),
            Mean = mean(age), Median = median(age),
            SD = round(sd(age),1), IQR = IQR(age))
```

TABLE 2.4: Summary of Age Distribution—COVID-19 Study

trt	n	Mean	Median	SD	IQR
PLA	10	34.0	33.5	3.1	3.50
TRT	10	33.6	33.5	4.1	4.75

The code below was used to create Figure 2.4, a histogram of **age**.

```
covid19_trial %>%
  ggplot(aes(age)) +
  geom_histogram(binwidth = 2, colour = "black", fill = "grey")
```

In Figure 2.4 the bin width is set to 2 via specifying the **binwidth** argument. This yields 7 rectangular bins (i.e., intervals) that have width equal to 2 and height equal to the number of values **age** that are contained in the bin.

The bins can also be used to construct rectangles with height y; y is called the density. The area of these rectangles is hy, where h is the bin width, and the sum of these rectangles is one. **geom_histogram(aes(y = ..density..))** will produce a histogram with density plotted on the vertical axis instead of count.

Another popular method to visualize the distribution of a continuous variable is the **boxplot** (see Figure 2.5). The box is composed of two "hinges" which represent the first and third quartile. The line in between the hinges represents

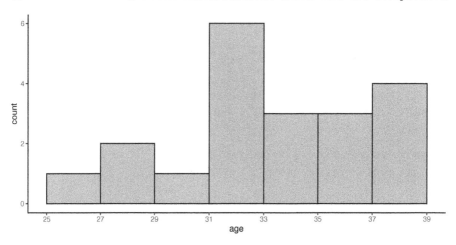

FIGURE 2.4: Distribution of Age—COVID-19 Study: Histogram

median age. The upper whisker extends from the hinge to the largest value no further than $1.5 * IQR$, and the lower whisker extends from the hinge to the smallest value no further than $1.5 * IQR$ where IQR is the inter-quartile range (75th percentile - 25th percentile).

```
covid19_trial %>% ggplot(aes(age, y = "")) +
  geom_boxplot() +
  scale_x_continuous(breaks = seq(min(covid19_trial$age),
                                  max(covid19_trial$age),
                                  by = 2)) +
  theme(
    axis.title.y = element_blank(),
    axis.ticks.y = element_blank(),
    axis.line.y = element_blank()
  )
```

2.3   Randomization

Two major applications of randomization are to: (1) select a representative sample from a population; and (2) assign "treatments" to subjects in experiments.

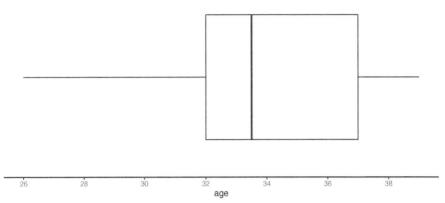

FIGURE 2.5: Distribution of Age—COVID-19 Study: Boxplot

2.3.1 Computation Lab: Randomization

Assume for the moment that the patients in Example 2.5 were randomly selected from, say, all people with COVID-19 in the United States during May, 2020. This means that each person with COVID-19 at this place and time had an equal probability of being selected (this was not the case in this study), and the data would be a random sample from the population of COVID-19 patients.

`covid_pats` is a data frame that contains the `id` of 25,000 COVID-19 patients. The code below can be used to randomly select 20 patients from `covid_pats`. A `tibble()` [Müller and Wickham, 2021] is an R data frame that works well with other **tidyverse** packages. In this book `tibble()` and `data.frame()` are mostly used interchangeably. Wickham and Grolemund [2016] discuss the differences between data frames and tibbles.

```
set.seed(101)
covid_sample <- tibble(id = sample(1:25000, 20))
```

A sample of 20 patients is drawn from the population of 25,000 patients using `sample()`. This is an example of simple random sampling: each patient has an equal chance of selection in the population not already selected.

When each patient agreed to enroll in the study, there was a 50-50 chance that they would receive a placebo, so chance decided what treatment a patient would receive, although neither the patient nor the researchers knew which treatment was assigned since the study was *double-blind*. Participants are to be randomized so that 10 receive active treatment `TRT` and 10 receive placebo `PLA`.

To assign treatments to subjects in `covid_sample`, `dplyr::sample_n()` randomly selects 10 rows from `covid_sample` and then creates a new variable called `treat` which is set equal to TRT.

```
set.seed(202)
trt_gp <- covid_sample %>%
  sample_n(10) %>%
  mutate(treat = "TRT")
```

The result can then be merged with `covid_sample`, where only rows *not in* `trt_gp` are included in the resulting merge via `dplyr::anti_join()`.

```
pla_gp <- covid_sample %>%
  anti_join(trt_gp) %>%
  mutate(treat = "PLA")
```

`dplyr::bind_rows()` is then used to merge the treatment `trt_gp` and placebo `pla_gp` groups, and the resulting data frame is stored in `covid_trtassign`.

```
covid_trtassign <- bind_rows(trt_gp,pla_gp)
```

The treatment assignments to four patients in the sample are shown in Table 2.5.

TABLE 2.5: Random Assignment of Treatment to Sample

id	treat
15203	TRT
3004	PLA
2873	PLA
13558	PLA

How many distinct ways are there to assign 10 subjects to TRT and 10 subjects to PLA? There are $\binom{20}{10}$ = 184,756 ways to assign 10 patients to TRT and 10 to PLA. We will return to this in later chapters.

2.4   Theoretical Distributions

The total aggregate of observations that *might* occur as a result of repeat-edly performing a particular study is called a **population** of observations. The observations that actually occur are a **sample** from this hypothetical population.

The distribution of continuous data can often be represented by a continuous curve called a **(probability) density function (p.d.f.)**, and the distribution of discrete data is represented by a **probability mass function**.

A **random variable** is essentially a random number—a function from the sample space to the real numbers. The probability measure on the sample space determines the probabilities of the random variable. Random variables will often be denoted by italic uppercase letters from the end of the alphabet. For example, Y might be the age of a subject participating in a study, and since the outcome of the study is random, Y is random. Observed data from a study will usually be denoted by lower case letters from the end of the alphabet. For example, y might be the observed value of a subject's age (e.g., $y = 32$).

If X is a discrete random variable, then the distribution of X is specified by its *probability mass function* $p(x_i) = P(X = x_i)$, $\sum_i p(x_i) = 1$. If X is a continuous random variable, then its distribution is fully characterized by its *probability density function* $f(x)$, where $f(x) \geq 0$, f is piecewise continuous, and $\int_{-\infty}^{\infty} f(x)dx = 1$.

The **cumulative distribution function (CDF)** of X is defined as follows:

$$F(x) = P(X \leq x) = \begin{cases} \int_{-\infty}^{x} f(x)dx & \text{if } X \text{ is continuous,} \\ \sum_{i=-\infty}^{x} p(x_i) & \text{if } X \text{ is discrete.} \end{cases}$$

If f is continuous at x, then $F'(x) = f(x)$ (fundamental theorem of calculus).

If X is continuous, then the CDF can be used to calculate the probability that X falls in the interval (a, b), and corresponds to the area under the density curve which can also be expressed in terms of the CDF:

$$P(a < X < b) = \int_{a}^{b} f(x)dx = F(b) - F(a).$$

The pth **quantile** of F is defined to be the value x_p such that $F(x_p) = p$

2.4.1   The Binomial Distribution

The binomial distribution models the number of "successes" in n independent trials, where the outcome of each trial is "success" or "failure". The terms "success" and "failure" can be any response that is binary, such as "yes" or "no", "true" or "false".

Let
$$X_i = \begin{cases} 1 & \text{if outcome is a success,} \\ 0 & \text{if outcome is a failure.} \end{cases}$$

The probability mass function is $P(X_i = 1) = p, i = 1, \ldots, n$. This is denoted as $X_i \sim Bin(n, p)$. When $n = 1$ the $Bin(1, p)$ is often called the Bernoulli distribution. The probability mass function is `dbinom()`, and the CDF is `pbinom()`. Figure 2.6 shows the binomial probability mass function of several different binomial distributions.

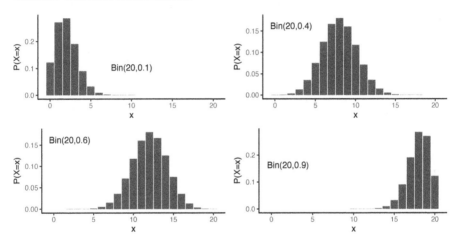

FIGURE 2.6: Binomial Probability Mass Function

2.4.2   The Normal Distribution

The density function of the normal distribution with mean μ and standard deviation σ is

$$\phi(x; \mu, \sigma) = \frac{1}{\sigma\sqrt{2\pi}} exp\left(\frac{-1}{2}\left(\frac{x-\mu}{\sigma}\right)^2\right)$$

The standard normal density is $\phi(x) = \phi(x; 0, 1)$ (Figure 2.7). The mean shifts the normal density along the horizontal axis, and the standard deviation scales the curve along the vertical axis.

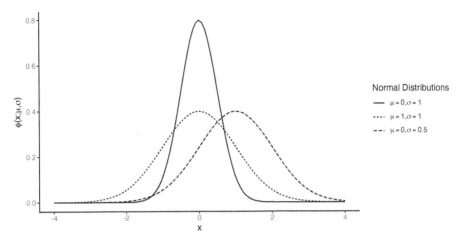

FIGURE 2.7: Normal Density Function

A random variable X that follows a normal distribution with mean $E(X) = \mu$ and variance $Var(X) = \sigma^2$ will be denoted by $X \sim N\left(\mu, \sigma^2\right)$.

$Y \sim N\left(\mu, \sigma^2\right)$ if and only if $Y = \mu + \sigma Z$, where $Z \sim N(0, 1)$.

The cumulative distribution function (CDF) of the $N(\mu, \sigma^2)$,

$$\Phi(x; \mu, \sigma^2) = P(X < x) = \int_{-\infty}^{x} \phi(u)du$$

is shown in 2.8 using the R function for the normal CDF `pnorm()` for various values of μ and σ.

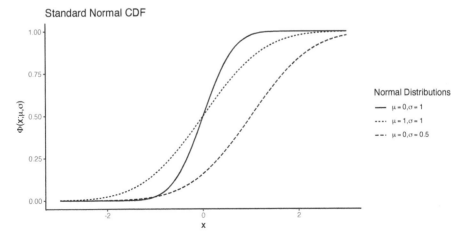

FIGURE 2.8: Normal Distribution Function

The CDF of the standard normal distribution will usually be denoted without μ and σ^2 as $\Phi(x)$.

Definition 2.1 (Normal Quantile). The $p^{th}, 0 < p < 1$ quantile of the $N(0,1)$ is the value z_p, such that $\Phi(z_p) = p$.

The density function ϕ of the $N(\mu, \sigma^2)$ distribution is symmetric about its mean μ, $\phi(\mu - x) = \phi(\mu + x)$.

Corollary 2.1 (Normal Quantile Symmetry). $z_{1-p} = -z_p$, $0 < p < 1$

Corollary 2.1 can be proved by noticing that the symmetry of ϕ implies that $1 - \Phi(\mu - x; \mu, \sigma^2) = \Phi(\mu + x; \mu, \sigma^2)$. Let $\mu = 0, \sigma^2 = 1$, and $x = z_p$ so $\Phi(z_p) + \Phi(-z_p) = 1$. $p = \Phi(z_p) \Rightarrow 1 - p = \Phi(-z_p)$, but $1 - p = \Phi(z_{1-p})$. So, $\Phi(-z_p) = \Phi(z_{1-p}) \Rightarrow z_{1-p} = -z_p$.

2.4.3　Chi-Square Distribution

Let $X_1, X_2, ..., X_n$ be i.i.d. $N(0,1)$. The distribution of

$$Y = \sum_{i=1}^{n} X_i^2,$$

has a chi-square distribution on n degrees of freedom or χ_n^2, denoted $Y \sim \chi_n^2$. $E(Y) = n$ and $Var(Y) = 2n$.

The chi-square distribution is a right-skewed distribution but becomes normal as the degrees of freedom increase. Figure 2.9 shows χ_{50}^2 density is very close to the $N(50, 100)$ density.

Let $X_1, X_2, ..., X_n$ be i.i.d. $N(0,1)$. The sample variance is $S^2 = \sum_{i=1}^{n} (X_i - \bar{X})^2 /(n - 1)$. The distribution of $\left(\frac{n-1}{\sigma^2}\right) S^2 \sim \chi_{n-1}^2$.

2.4.4　t Distribution

Let $X \sim N(0,1)$ and $W \sim \chi_n^2$ be independent. The distribution of $\frac{X}{\sqrt{W/n}}$ has a t distribution on n degrees of freedom or $\frac{X}{\sqrt{W/n}} \sim t_n$.

Let $X_1, X_2, ..., X_n$ be i.i.d. $N(0,1)$. The distribution of $\sqrt{n}\left(\bar{X} - \mu\right)/S \sim t_{n-1}$, where S^2 is the sample variance and $\bar{X} = n^{-1} \sum_{i=1}^{n} X_i$. This follows since $\bar{X}$ and S^2 are independent.

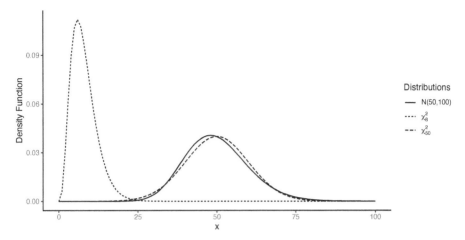

FIGURE 2.9: Comparison of Chi-Square Distributions

The t distribution for small values of n has "heavier tails" compared to the normal. As the degrees of freedom increase the t-distribution is almost identical to the normal distribution.

Figure 2.10 shows the density for various degrees of freedom. As the degrees of freedom becomes larger, the distribution is closer to the normal distribution, and for smaller degrees of freedom there is more probability in the tails of the distribution.

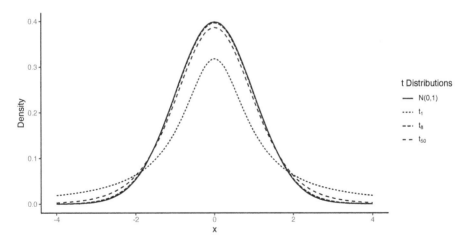

FIGURE 2.10: Comparison of t Distributions

2.4.5   F Distribution

Let $X \sim \chi_m^2$ and $Y \sim \chi_n^2$ be independent. The distribution of

$$W = \frac{X/m}{Y/n} \sim F_{m,n},$$

has an F distribution on m, n degrees of freedom, denoted $F_{m,n}$. The F distribution is a right skewed distribution (see Figure 2.11).

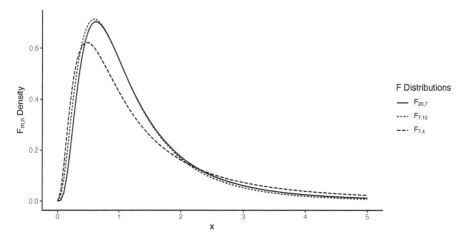

FIGURE 2.11: Comparison of F Distributions

2.4.6   Computation Lab: Theoretical Distributions

2.4.6.1   Binomial Distribution

In R a list of all the common distributions can be obtained by the command `help("distributions")`.

Example 2.6. Use R to calculate the $P(4 \le X < 7)$ and $P(X > 8)$ where $X \sim Bin(10, 0.2)$.

$P(4 \le X < 7) = P(X \le 6) - P(X \le 4) = F(6) - F(4)$, where $F(\cdot)$ is the CDF of the binomial distribution. `pbinom()` can be used to calculate this probability.

```
pbinom(q = 6, size = 10, prob = 0.2) - pbinom(q = 4, size = 10,
    prob = 0.2)
```

```
## [1] 0.03193
```

2.4.6.2 Normal Distribution

Example 2.7. Use R to calculate $P(X > 5)$, where $X \sim N(6, 2)$.

$P(X > 5) = 1 - P(X < 5)$. `pnorm()` can be used to calculate this probability.

```
1 - pnorm(5, mean = 6, sd = sqrt(2))
```

```
## [1] 0.7602
```

Example 2.8. Use R to show that $-z_{0.975} = z_{1-0.975}$.

`qnorm()` can be used to calculate $z_{0.975}$, and $z_{1-0.975}$.

```
-1*qnorm(p = 0.975) == qnorm(p = 1 - 0.975)
```

```
## [1] TRUE
```

2.4.6.3 Chi-Square Distribution

Example 2.9. If $X_1, X_2, ..., X_{50}$ are a random sample from a $N(40, 25)$ then calculate $P(S^2 > 27)$.

We know that $\frac{(50-1)}{25}S^2 \sim \chi^2_{(50-1)}$. So

$$P(S^2 > 27) = P\left(\frac{49}{25}S^2 > \frac{49}{25} \times 27\right)$$
$$= P\left(\chi^2_{49} > 52.92\right).$$

The CDF of the χ^2_n, $P(\chi^2_n < q)$ in R is `pchisq()`.

```
1 - pchisq(q = 52.92, df = 49)
```

```
## [1] 0.3253
```

So, $P(S^2 > 27) = 1 - P(\chi^2_{49} < 52.92) = 0.3253$.

2.4.6.4   t Distribution

Example 2.10. A random sample of body weights (kg) from 10 male subjects are 66.9, 70.9, 65.8, 78.0, 71.6, 65.9, 72.4, 73.7, 72.9, and 68.5. The distribution of weight in this population is known to be normally distributed with mean 70kg. What is the probability that the sample mean weight is between 68kg and 71kg?

The distribution of $\sqrt{10}\left(\bar{X} - 70\right)/S \sim t_9$, where S is the sample standard deviation. The sample standard deviation can be calculated using R:

```
weight_data <- c(66.9, 70.9, 65.8, 78.0, 71.6,
                 65.9, 72.4, 73.7, 72.9, 68.5)
sd(weight_data)
```

```
[1] 3.904
```

So

$$P\left(68 < \bar{X} < 71\right) = P\left(\frac{68 - 70}{3.904/\sqrt{10}} < \frac{\bar{X} - 70}{3.904/\sqrt{10}} < \frac{71 - 70}{3.904/\sqrt{10}}\right).$$

Use `pt()` the CDF of the t_n distribution in R to calculate the probability.

```
low_lim <- (68-70)/(sd(weight_data)/sqrt(10))
up_lim <- (71-70)/ (sd(weight_data)/sqrt(10))
pt(up_lim, df = 9) - pt(low_lim, df = 9)
```

```
[1] 0.7107
```

$$P\left(68 < \bar{X} < 71\right) = P\left(-1.62 < t_9 < 0.81\right) = 0.7107.$$

2.4.6.5   F Distribution

Example 2.11. Consider Example 2.10. Another random sample of 20 subjects from the same population is obtained. Let S_1^2 be the sample variance of the first sample and S_2^2 be the sample variance of the second sample. Use R to find the probability that the ratio of the second sample variance to the first sample variance exceeds 5.

We are asked to calculate $P\left(S_2^2/S_1^2 > 5\right)$. The first and second samples are independent since we are told it's a separate sample from the same population, and the distribution of weights in this sample will also be $N(70, \sigma^2)$. Now, we know that $(20-1)/\sigma^2 S_2^2 \sim \chi^2_{(20-1)} \Rightarrow S_2^2/\sigma^2 \sim \chi^2_{19}/19$, and $(10-1)/\sigma^2 S_1^2 \sim \chi^2_{(10-1)} \Rightarrow S_1^2/\sigma^2 \sim \chi^2_9/9$.

$$
\begin{aligned}
P\left(S_2^2/S_1^2 > 5\right) &= P\left((S_2^2/\sigma^2)/(S_1^2/\sigma^2) > 5\right) \\
&= P\left(((\chi^2_{19}/19)/(\chi^2_9/9) > 5\right) \\
&= P\left(F_{19,9} > 5\right)
\end{aligned}
$$

Use `pf()` the CDF of the F_{df_1, df_2} distribution in R to calculate the probability.

```
1- pf(q = 5, df1 = 19, df2 = 9)
```

```
## [1] 0.008872
```

$P\left(S_2^2/S_1^2 > 5\right) = 0.0089$. It's unlikely that the sample variance of another sample from the same population would be five times the variance of the first sample.

2.5 Quantile-Quantile Plots

Quantile-quantile (Q-Q) plots are useful for comparing distribution functions. If X is a continuous random variable with strictly increasing distribution function $F(x)$ then the *pth* quantile of the distribution is the value x_p such that,

$$
F(x_p) = p \text{ or } x_p = F^{-1}(p).
$$

In a Q-Q plot, the quantiles of one distribution are plotted against another the quantiles of another distribution. Q-Q plots can be used to investigate whether a set of numbers follows a certain distribution.

Consider independent observations $X_1, X_2, ..., X_n$ from a uniform distribution on $[0, 1]$ or Unif$[0, 1]$. The ordered sample values (also called the order statistics) are the values $X_{(j)}$ such that $X_{(1)} < X_{(2)} < \cdots < X_{(n)}$.

It can be shown that $E\left(X_{(j)}\right) = \frac{j}{n+1}$. This suggests that if the underlying distribution is Unif[0,1] then a plot of $X_{(j)}$ vs. $\frac{j}{n+1}$ should be roughly linear.

Figure 2.12 shows two random samples of 1,000 from Unif$[0, 1]$ and $N(0, 1)$ plotted against their order, $j/(n+1)$, with a straight line to help assess linearity. The points from the Unif$[0, 1]$ follow the reference line, while the points from the $N(0, 1)$ *systematically* deviate from this line.

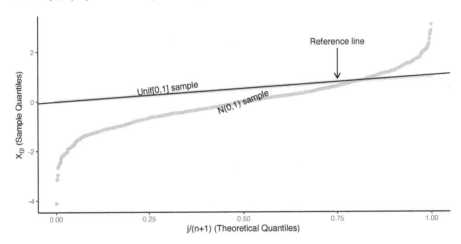

FIGURE 2.12: Expected Uniform Order Statistics

The reference line in Figure 2.12 is a line that connects the first and third quartiles of the Unif$[0, 1]$. Let x_{25} and x_{75} be the first and third quartiles of the uniform distribution, and y_{75} and y_{25} be the first and third quartiles of the ordered sample. The reference line in Figure 2.12 is the straight line that passes through (x_{25}, y_{25}) and (x_{75}, y_{75}).

A continuous random variable with strictly increasing CDF F_X can be transformed to a Unif$[0, 1]$ by defining a new random variable $Y = F_X(X)$. This transformation is called the *probability integral transformation.*

This suggests the following procedure. Suppose that it's hypothesized that a sample $X_1, X_2, ..., X_n$ follows a certain distribution, with CDF F. We can plot $F(X_{(j)})$ vs. $\frac{j}{n+1}$, or equivalently

$$X_{(j)} \quad \text{vs.} \quad F^{-1}\left(\frac{j}{n+1}\right) \tag{2.1}$$

$X_{(j)}$ can be thought of as empirical quantiles and $F^{-1}\left(\frac{j}{n+1}\right)$ as the hypothesized quantiles.

There are several reasonable choices for assigning a quantile to $X_{(j)}$. Instead of assigning it $\frac{j}{n+1}$, it is often assigned $\frac{j-0.5}{n}$. In practice it makes little difference which definition is used.

Figure 2.13 shows two ordered random samples of 1000 from the Unif$[0, 1]$ and $N(0, 1)$ plotted against $\Phi^{-1}(k/(n + 1))$ (see (2.1)). The points from the Unif$[0, 1]$ systematically deviate from the reference line.

2.5.1 Computation Lab: Quantile-Quantile Plots

The R code below can be used to produce a plot similar to Figure 2.13, but without annotations and transparency to show the different samples. $\Phi^{-1}(x)$ can be computed using qnorm(). rnorm() and runif() simulate random numbers from the normal and uniform distributions, and geom_abline(slope = b, intercept = a) adds the line y=a+bx to the plot.

```
n <- 1000
y <- sort(rnorm(1000))
q <- quantile(y, c(0.25,0.75))
slope <- (q[2]-q[1])/(qnorm(.75)-qnorm(.25))
int <- q[1]-slope*qnorm(.25)

tibble(y = sort(rnorm(n)), x = qnorm(1:n/(n+1))) %>%
  ggplot(aes(x,y)) + geom_point() +
  geom_point(data = tibble(y = sort(runif(n)),
                           x = qnorm(1:n/(n+1)))) +
  geom_abline(slope = slope, intercept = int)
```

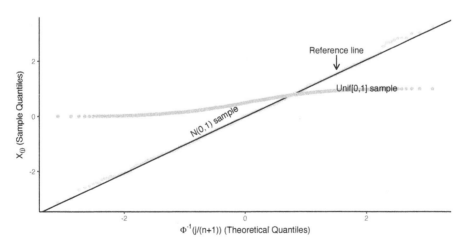

FIGURE 2.13: Comparing Sample Quantiles with Normal Distribution

Quantile-quantile plots are implemented in many graphics libraries including ggplot2. geom_qq() produces a plot of the sample versus the theoretical quantiles, and geom_qq_line() adds a reference line to assess if the points follow a linear pattern. The default distribution in `geom_qq()` and `geom_qq_line()` is the standard normal, but it's possible to specify another distribution using the `distribution` and `dparms` parameters.

Figure 2.14 shows the Q-Q plots comparing a simulated sample from a χ^2_8 to a $N(0,1)$ (left plot) and, the correct, χ^2_8 (right plot).

```
set.seed(27)
y <- rchisq(n = 1000, df = 8)

p1 <- tibble(y) %>%
  ggplot(aes(sample = y)) +
  geom_qq() +
  geom_qq_line() +
  ylab("Sample Quantiles") +
  xlab("Theoretical Quantiles")

p2 <- tibble(y = y) %>%
  ggplot(aes(sample = y)) +
  geom_qq(distribution = qchisq, dparams = list(df = 8)) +
  geom_qq_line(distribution = qchisq, dparams = list(df = 8)) +
  ylab("Sample Quantiles") +
  xlab("Theoretical Quantiles")

p1 + p2
```

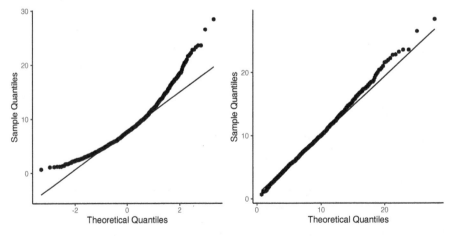

FIGURE 2.14: Q-Q Plots using geom_qq()

Example 2.12. agedat contains the ages of participants in a study of social media habits. Is it plausible to assume that the data are normally distributed?

A summary of the distribution is

```
agedata %>%
  summarise(N = n(),
            Mean = mean(age),
            Median = median(age),
            SD = sd(age),
            IQR = IQR(age))
```

```
# A tibble: 1 x 5
      N  Mean Median    SD   IQR
  <int> <dbl>  <dbl> <dbl> <dbl>
1    50  36.7     35  6.88  2.75
```

The mean and median are very close, but the standard deviation is much larger than the IQR, and the left panel of Figure 2.15 shows the distribution of age is symmetric.

Neither the distribution summary nor the left panel of Figure 2.15 allows us to evaluate whether the data is normally distributed.

If the data has a normal distribution, then it should be close to a $N(37, 49)$. A Q-Q plot to assess this is shown in the right panel of Figure 2.15.

```
p1 <- agedata %>% ggplot(aes(x = age)) +
  geom_histogram(bins = 10, colour = "black", fill = "grey") +
  ggtitle("Distribution of Age")

p2 <- agedata %>%
  ggplot(aes(sample = age)) +
  geom_qq(dparams = list(mean = 37, sd = 7)) +
  geom_qq_line(dparams = list(mean = 37, sd = 7)) +
  ggtitle("Normal Quantile Plot of Age") +
  xlab("Theoretical Quantiles") +
  ylab("Sample Quantiles")

p1 + p2
```

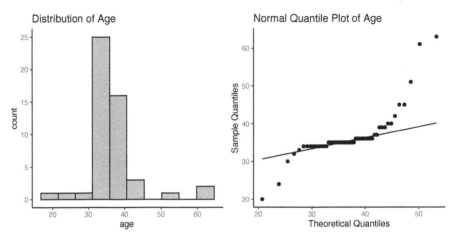

FIGURE 2.15: Age Plots

The histogram and the Q-Q plot indicate the tails of the age distribution are heavier than the normal distribution. The right panel of Figure 2.15 shows a systematic deviation from the reference line for very small and large sample quantiles, which implies that the data are not normally distributed.

2.6   Central Limit Theorem

Theorem 2.1 (Central limit theorem). *Let $X_1, X_2, \ldots$ be an i.i.d. sequence of random variables with mean $\mu = E(X_i)$ and variance $\sigma^2 = Var(X_i)$.*

$$\lim_{n \to \infty} P\left(\frac{\bar{X} - \mu}{\frac{\sigma}{\sqrt{n}}} \le x \right) = \Phi(x),$$

where $\bar{X} = \sum_{i=1}^{n} X_i / n$.

The central limit theorem states that the distribution of $\bar{X}$ is approximately $N\left(\mu, \sigma^2/n\right)$ for large enough n, provided the X_i's have finite mean and variance.

2.6.1   Computation Lab: Central Limit Theorem

Example 2.13. Seven teaching hospitals affiliated with the University of Toronto delivered 33,481 babies between July 1, 2019 and June 31, 2020 [Obstetrics and Gynaecology, 2019]. What is the distribution of the average number of female births at University of Toronto teaching hospitals?

Let $X_i = 1$, if a baby is born female, and $X_i = 0$ if the baby is male, $i = 1, \ldots, 33481$. The probability of a female birth is 0.5, $P(X_i = 1) = 0.5$, and $X_i \sim Bin(1, 0.5)$.

The average number of females is $\bar{X} = \sum_{i=1}^{33481} X_i / 33481$. $E(X_i) = 0.5$ and $Var(X_i) = p(1 - p) = 0.5(1 - 0.5) = 0.25$.

$$\frac{\sqrt{33481}\left(\bar{X} - 0.5\right)}{0.25} \overset{approx}{\underset{\sim}{}} N(0, 1).$$

This is known as the sampling distribution of $\bar{X}$.

A simulation of size 100 (`N <- 100`) from the sampling distribution of $\bar{X}$, with $X_i \sim Bin(n, p)$, where $n = 1, p = 0.5$ is implemented below. Each simulated data set has 33,481 rows (`M <- 33481`) representing a hypothetical data set of births for the year. `sprintf("SIM%d",1:N)` renames the columns of `sims` to SIM1, SIM2,...,SIM100.

```
set.seed(28)
N <- 100
M <- 33481
n <- 1
p <- 0.5
sims <- replicate(N, rbinom(M, n, p))
colnames(sims) <- sprintf("SIM%d", 1:N)
female_babies <- as_tibble(sims)

head(female_babies, n = 3)
```

```
## # A tibble: 3 x 100
##    SIM1  SIM2  SIM3  SIM4  SIM5  SIM6  SIM7  SIM8  SIM9
##   <int> <int> <int> <int> <int> <int> <int> <int> <int>
## 1    0     1     1     0     0     0     0     0     0
## 2    0     1     1     1     1     0     0     1     1
## 3    0     0     0     0     0     0     1     0     0
## # ... with 91 more variables: SIM10 <int>,
## #    SIM11 <int>, SIM12 <int>, SIM13 <int>,
## #    SIM14 <int>, SIM15 <int>, SIM16 <int>,
## #    SIM17 <int>, SIM18 <int>, SIM19 <int>,
## #    SIM20 <int>, SIM21 <int>, SIM22 <int>,
## #    SIM23 <int>, SIM24 <int>, SIM25 <int>,
## #    SIM26 <int>, SIM27 <int>, SIM28 <int>, ...
```

The R code to generate Figure 2.16 is shown below.

```
female_babies %>%
  summarise(across(all_of(colnames(female_babies)), mean)) %>%
  pivot_longer(
    cols = colnames(female_babies),
    names_to = "sim",
    values_to = "mean"
  ) %>%
  ggplot(aes(sample = mean)) +
  geom_qq() +
  geom_qq_line()
```

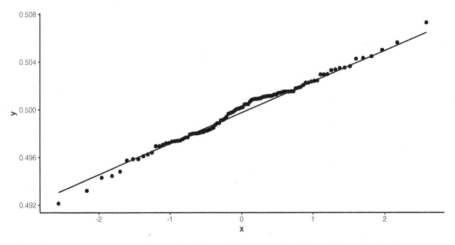

FIGURE 2.16: Normal Quantile Plot of Average Number of Female Babies

The figure shows that the distribution of average number of females born is normally distributed. The mean of each column was computed using **across()** giving a data frame with one row and one hundred columns. **ggplot()** requires that the data be shaped so that the columns correspond to the variables that are to be plotted, and the rows represent the observations. **tidyr::pivot_longer()** function reshapes the data into this format: The **names_to** parameter names the new column **sim** where the values of the rows will be SIM1 to SIM100; **values_to** parameter will create a column **mean** where each row contains the mean value of a simulation.

2.7 Statistical Inference

2.7.1 Hypothesis Testing

The hypothesis testing framework is a method to infer plausible parameter values from a data set.

Example 2.14. Let's continue with Example 2.13. Investigators want to compare the Caesarean section (C-section) rate between Michael Garrison Hospital (MG) and North York General Hospital (NYG). Table 2.6 summarises the data. Is the C-section rate in the two hospitals statistically different?

TABLE 2.6: C-Section Rates by Hospital

Hospital	Total Births	C-Section Rate
MG	2840	29.0%
NYG	4939	32.0%

The *observed* difference in the proportions of C-section rates is calculated as the difference between $\hat{p}_{MG}$, and $\hat{p}_{NYG}$—the proportion of C-sections at MG and NYG: $\hat{p}_{MG} - \hat{p}_{NYG}$ = -0.03.

2.7.2 Hypothesis Testing via Simulation

How likely are we to observe a difference of this or greater magnitude, **if the proportions of C-sections in the two hospitals are equal**? If it's unlikely, then this is evidence that either (1) the assumption that the proportions are equal is false or (2) a rare event has occurred.

Assume that MG has 2840 births per year, and NYG has 4939 births per year, but the probability of a C-section at both hospitals is the same. The latter statement will correspond to the hypothesis of no difference or the *null hypothesis*. The null hypothesis H_0 can be formally stated as $H_0 : p_{MG} = p_{NYG}$, where p_{MG} and p_{NYG} are the probabilities of a C-section at MG and NYG.

If $H_0 : p_{MG} = p_{NYG}$ is true, then

1. A woman who gives birth at MG or NYG should have the same probability of a C-section.

2. If 2,840 of the 7,779 women are randomly selected and labeled as MG and the remaining 4,939 women are labeled as NYG then the proportions of C-sections $\hat{p}_{MG}^{(1)}$ and $\hat{p}_{NYG}^{(1)}$ should be almost identical or $\hat{p}_{MG}^{(1)} - \hat{p}_{NYG}^{(1)} \approx 0$.

3. If the random selection in 2. is repeated a large number of times, say N, and the differences for the ith random selection $\hat{p}_{MG}^{(i)} - \hat{p}_{NYG}^{(i)}$ are calculated each time, then this will approximate the sampling distribution of $\hat{p}_{MG} - \hat{p}_{NYG}$ (assuming H_0 is true). $\hat{p}_{MG}^{(i)}$ and $\hat{p}_{NYG}^{(i)}$ are the proportions of C-sections in MG and NYG after the hospital labels MG and NYG have been randomly shuffled. In order to randomly shuffle the hospital labels we can randomly select 2,840 observations from the data set and label them MG, and label the remaining 7,779-2,840=4,939 observations as NYG. Remember if H_0 is true, then it shouldn't matter which 2,840 observations are labeled as MG and which 4,939 are labeled as NYG.

2.7.3   Model-based Hypothesis Testing

Another method to evaluate the question in Example 2.14 is to assume that the distribution of C-sections within each hospital follows a binomial distribution. If H_0 is true, then the theoretical sampling distribution of $\hat{p}_{MG} - \hat{p}_{NYG}$ is $N(0, p(1 - p)(1/2840 + 1/4939))$, where p is the probability of a C-section. We can evaluate H_0 by testing whether the mean of this normal distribution is zero. The null hypothesis does not specify a value of p, but is usually estimated using the combined sample to estimate the proportion:

The normal deviate is

$$z = \frac{\hat{p}_{MG} - \hat{p}_{NYH}}{\sqrt{\hat{p}(1 - \hat{p})(1/n_1 + 1/n_2)}} = \frac{-0.03}{\sqrt{0.31(0.69)(1/2840 + 1/4939)}} = -2.75,$$

so, the p-value for a two-sided test is $P(|Z| > |-2.75|) = 2 * \Phi(-2.75) = 0.01$

2.7.4   Computation Lab: Hypothesis Testing

Now, let's implement the *simulation* described in 2.7.2.

```
set.seed(9)

N <- 1000
samp_dist <- numeric(N)        # 1. output
for (i in seq_along(1:N)) {    # 2. sequence
  shuffle <- sample(x = 7779, size = 2840) # 3. body
  p_MG_shuffle <- sum(CSectdat$CSecthosp[shuffle]) / 2840
  p_NYG_shuffle <- sum(CSectdat$CSecthosp[-shuffle]) / 4939
  samp_dist[i] <- p_MG_shuffle - p_NYG_shuffle
}
```

This simulation code has three components:

1. output: `samp_dist <- numeric(N)` allocates a numeric vector of size N to store the output.

2. sequence: `i in seq_along(1:N)` determines what to loop over: each iteration will assign i to a different value from the sequence `1:N`.

3. body: `shuffle <- sample(x = 7779, size = 2840)` selects a random sample of `size = 2840` from the sequence `1:7779` and stores the result in `shuffle`, `p_MG_shuffle` and `p_NYG_shuffle` are the shuffled proportions of C-sections in the MG and NYG, and `samp_dist[i]` stores the difference in shuffled proportions for the ith random shuffle.

Figure 2.17 shows a histogram of the sampling distribution (also see Section 3.7) of the test statistic $\hat{\delta} = \hat{p}_{MG} - \hat{p}_{NYG}$. The distribution is symmetric and centred around zero.

The R code for Figure 2.17 without annotations is shown below.

```
tibble(samp_dist) %>%
  ggplot(aes(samp_dist)) +
  geom_histogram(bins = 35,
                 colour = "black",
                 fill = "grey")
```

Figure 2.18 shows that the simulated sampling distribution is normally distributed even though there are no assumptions about the distribution of $\hat{\delta}$.

```
tibble(samp_dist) %>%
  ggplot(aes(sample = samp_dist)) +
  geom_qq() +
  geom_qq_line()
```

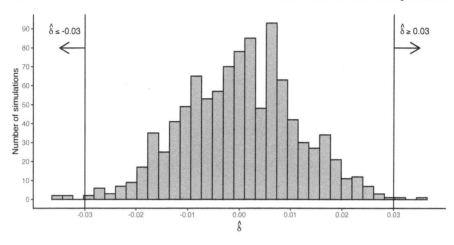

FIGURE 2.17: Sampling Distribution of Difference of Proportions

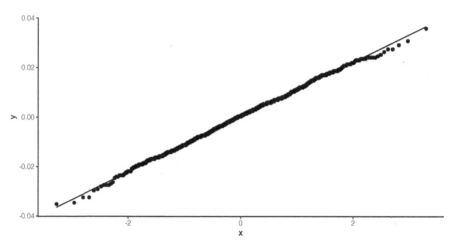

FIGURE 2.18: Normal Q-Q Plot for Sampling Distribution of Difference in C-Section Proportions

The proportion of simulated differences that are as extreme or more extreme (in either the positive or negative direction) than the observed difference of -0.03 is

```
p_value_sim <- sum(samp_dist >= abs(obs_diff) |
                   samp_dist <= -1*abs(obs_diff))/N
p_value_sim
```

```
## [1] 0.006
```

`p_value_sim` is called the two-sided *p-value* of the test. Six of the 1000 simulations were as extreme or more extreme than the observed difference, assuming H_0 is true (see Figure 2.17). This means that either a rare event (i.e., the event of observing a test statistic as extreme or more extreme than the observed value) has occurred or the assumption that H_0 is true is **not** supported by the data. This test, or any other hypothesis tests, can never say definitively which is the case, but it's common practice to interpret a p-value this small as *weak* evidence to support the assumption that H_0 is true. In other words, there is strong evidence that C-section rates in the two hospitals are different.

Alternatively, we could use R to conduct a model-based hypothesis test comparing the proportions of C-sections in Example 2.14.

```
CSectdat %>% summarise(Births = n(),
                       Total = sum(CSecthosp),
                       Rate = Total/Births)
```

```
## # A tibble: 1 x 3
##    Births Total   Rate
##     <int> <int>  <dbl>
## 1    7779  2376  0.305
```

`stats::prop.test()` can be used to calculate the p-value for a model-based hypothesis test.

```
two_prop_test <- prop.test(n = c(2840,4939), x = c(819,1557),
                           correct = F)
two_prop_test$p.value
```

```
## [1] 0.01326
```

The p-values from the simulation and model-based tests both yield strong evidence that C-section rates in the two hospitals are different.

2.7.5   Confidence Intervals

Consider the mean $\bar{X}$ of a sample $X_1, \ldots, X_n$, where the X_i are i.i.d. with mean μ and variance σ^2. The central limit theorem tells us that the distribution of $\bar{X}$ is approximately $N(\mu, \sigma^2/n)$. This means that apart from a 5% chance

$$\mu - 1.96\sigma/\sqrt{n} < \bar{X} < \mu + 1.96\sigma/\sqrt{n}.$$

The left-hand inequality is equivalent to $\mu < \bar{X} + 1.96\sigma/\sqrt{n}$, and the right-hand inequality is the same as $\mu > \bar{X} - 1.96\sigma/\sqrt{n}$. Together these two statements imply $\bar{X} - 1.96\sigma/\sqrt{n} < \mu < \bar{X} + 1.96\sigma/\sqrt{n}$, apart from a 5% chance when the sample was drawn.

The uncertainty attached to a confidence interval for μ comes from the variability in the sampling process. We don't know if any particular interval contains μ or not. In the long run (i.e., in repeated sampling), 95% of the intervals will include μ. For example, if the study was replicated one thousand times and a 95% confidence interval was calculated for each study, then fifty confidence intervals would not contain μ.

$$\bar{X} - (z_{1-\alpha/2})\sigma/\sqrt{n} < \mu < \bar{X} + (z_{1-\alpha/2})\sigma/\sqrt{n},$$

is a $100(1-\alpha)\%$ confidence interval for μ.

2.7.6   Computation Lab: Confidence Intervals

The simulation below illustrates the meaning behind "95% of intervals will include the *true* μ." The simulation generates N data sets of size n from the $N(0,1)$ distribution. If a confidence interval is calculated for each of the N data sets, then how many will contain the true $\mu = 0$?

```
set.seed(25) # 1. Setup parameters
N <- 1000
n <- 50
alpha <- 0.05
confquant <- qnorm(p = 1- alpha/2)

covered <- logical(N) #2. Output

for (i in seq_along(1:N)){ # 3. Sequence
  normdat <- rnorm(n)        # 4. Body
  ci.low <- mean(normdat) - confquant*1/sqrt(n)
  ci.high <- mean(normdat) + confquant*1/sqrt(n)

  if (ci.low < 0 & ci.high > 0) {
    covered[i] <- TRUE
    } else {
      covered[i] <- FALSE
    }
  }
sum(covered)


## [1] 947
```

947 of the 1000 simulations contain or cover the true mean, slightly less than 950 predicted by theory, but for practical purposes close enough to 950.

1. Setup parameters: N is the number of simulations; n is the sample size; `alpha` is the (type I - see Section 2.7.7) error rate for the confidence interval; and `confquant <- qnorm(p = 1- alpha/2)` is the normal quantile associated with the confidence interval.

2. Output: `covered` is a logical vector that will store Boolean values to indicate if the confidence interval contains zero.

3. Sequence: `i in seq_along(1:N)` determines what to loop over: each iteration will assign `i` to a different value from the sequence `1:N`.

4. Body: `rnorm(n)` generates a sample of size n from the $N(0,1)$; `ci.low` and `ci.high` calculate the lower and upper limits of the confidence interval; if `(ci.low < 0 & ci.high > 0)` is true, then the ith confidence interval contains zero and `covered[i]` is TRUE; otherwise it's FALSE.

2.7.7   Errors in Hypothesis Testing

In hypothesis testing there are two types of errors that can be made. They are called type I and type II errors.

	H_0 true	H_a true
Accept H_0	correct decision	type II error
Reject H_0	type I error	correct decision

The probabilities of type I and II errors are usually set as part of a study design.

$$\alpha = P(\text{type I error}), \ \beta = P(\text{type II error}).$$

If the p-value $\leq \alpha$, then the test is **statistically significant** at level α. The power of the test is $1 - \beta$, the probability of rejecting H_0 when the alternative hypothesis H_1 is true. A feasible way to control power is through sample size (see Chapter 4).

2.8   Linear Regression

2.8.1   Simple Linear Regression

Simple linear regression models the relationship between a quantitative variable, y, as a linear function of a quantitative variable, x.

Example 2.15 (World Happiness Report). The World Happiness Report [Helliwell et al., 2019] collects data from the Gallup World Poll global survey that measures life satisfaction using a "Cantril Ladder" question: think of a ladder with the best possible life being a 10 and the worst possible life being a 0.

Figure 2.19 shows the relationship between life satisfaction and childhood mortality in 2017 after a logarithmic transformation.

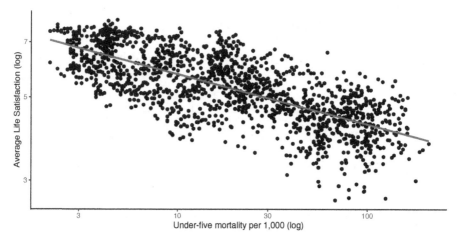

FIGURE 2.19: Life Satisfaction versus Child Mortality in 2017 (Example 2.15)

A **simple linear regression model** of life satisfaction versus childhood mortality is fit by estimating the intercept and slope in the equation:

$$y_i = \beta_0 + \beta_1 x_i + \epsilon_i, i = 1, ..., n \tag{2.2}$$

In this example, y_i and x_i are the ith values of the logarithmic values of life satisfaction and childhood mortality.

Least-squares regression minimizes $\sum_{i=1}^{n} \epsilon_i^2$ in (2.2). The values of $\hat{\beta}_0$ and $\hat{\beta}_1$ that minimize (2.2) are called the **least squares estimators**. $\hat{\beta}_0 = \bar{y} - \hat{\beta}_1 \bar{x}$,

$\hat{\beta}_1 = r\frac{S_y}{S_x}$, where r is the correlation between y and x, and S_x and S_y are the sample standard deviations of x and y respectively.

The ith **predicted value** is $\hat{y}_i = \hat{\beta}_0 + \hat{\beta}_1 x_i$, and the ith **residual** is $e_i = y_i - \hat{y}_i$.

Least squares works *well* when the **Gauss-Markov conditions** hold in (2.2):

1. $E(\epsilon_i) = 0$, for all i.
2. $Var(\epsilon_i) = \sigma^2$.
3. $E(\epsilon_i \epsilon_j) = 0$, for all $i \neq j$.

1 excludes data where a straight line is inappropriate; 2 avoids data where the variance is non-constant (this is often called heteroscedasticity); and 3 requires observations to be uncorrelated.

One popular measure of fit for simple linear regression models is

$$R^2 = 1 - \sum_{i=1}^{n} e_i^2 / \sum_{i=1}^{n} (y_i - \bar{y})^2.$$

R^2 is the square of the correlation coefficient between y and x, and has a useful interpretation as the proportion of variation that is explained by the linear regression of y on x for simple linear regression.

If $\epsilon_i \sim N(0, \sigma^2)$, then
$$y_i \sim N(\beta_0 + \beta_1 x_i, \sigma^2).$$

It can be shown that

$$(\hat{\beta}_j - \beta_j)/se(\hat{\beta}_j) \sim t_{n-2},$$

for regression models with $\beta_0 \neq 0$. This can be used to form tests and confidence intervals for the $\beta's$.

2.8.2   Multiple Linear Regression

If there are $k > 1$ independent variables, then (2.2) can be extended to a multiple linear regression model.

$$y_i = \beta_0 + \beta_1 x_{i1} + \beta_2 x_{i2} + \cdots + \beta_k x_{ik} + \epsilon_i, \ i = 1, ..., n, \tag{2.3}$$

where $Var(\epsilon_i) = \sigma^2$, $i = 1, ..., n$.

This can also be expressed in matrix notation as

$$\underset{n\times 1}{\mathbf{y}} = \underset{n\times k}{X}\underset{n\times 1}{\boldsymbol{\beta}} + \underset{n\times 1}{\boldsymbol{\epsilon}}. \qquad (2.4)$$

The least squares estimator can then be expressed as

$$\hat{\boldsymbol{\beta}} = \left(X'X\right)^{-1} X'\mathbf{y}, \qquad (2.5)$$

and the covariance matrix of $\hat{\beta}$ is

$$\mathrm{cov}\left(\hat{\beta}\right) = \left(X'X\right)^{-1}\sigma^2. \qquad (2.6)$$

An estimator of σ^2 is $\hat{\sigma}^2 = \frac{1}{n-k}\sum_{i=1}^{n}(y_i - \hat{y}_i)^2$, where $\hat{y}_i = \hat{\beta}_0 + \hat{\beta}_1 x_{i1} + \cdots + \hat{\beta}_k x_{ik}$ is the predicted value of y_i.

2.8.3   Categorical Covariates in Regression

Covariates x_{ij} in (2.3) can either be continuous or categorical variables. If a categorical independent variable has k categories then $k-1$ indicator variables should be used to index the categories provided the regression model contains a slope. If the regression model does not contain a slope, then exactly k indicator variables are required.

Investigators are interested in the relationship between Life Satisfaction and Region of the world in 2017 (Example 2.15). Figure 2.20 shows that satisfaction is highest in North America and lowest in Sub-Saharan Africa and South Asia. This relationship can be modelled using linear regression with a categorical covariate.

$$y_i = \beta_0 + \sum_{k=1}^{6} \beta_k x_{ik} + \epsilon_i, \qquad (2.7)$$

where

$$x_{ik} = \begin{cases} 1 & \text{if Country } i \text{ belongs to Region } k \\ 0 & \text{Otherwise.} \end{cases}$$

In this case the covariate Region has seven unordered categories: East Asia & Pacific, Europe & Central Asia, Latin America & Caribbean, Middle East & North Africa, North America, South Asia, and Sub-Saharan Africa.

$$E(y_i) = \beta_0 + \sum_{k=1}^{6} \beta_k x_{ik}.$$

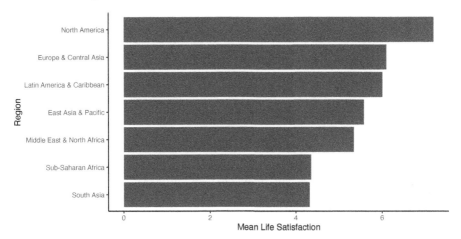

FIGURE 2.20: Mean Life Satisfaction by Region in 2017

If $k = 1, \ldots, 6$, then $\mu_k = E(y_i) = \beta_0 + \beta_k$, and if $k = 0$, then $\mu_0 = E(y_i) = \beta_0$ then $\beta_k = \mu_k - \mu_0$. In this case μ_0 is the mean of the reference category. So, $\beta_k, k = 1, \ldots, 6$ is the mean difference between the kth category and the reference category.

Example 2.16 (Hotelling's Weighing Problem). Hotelling [1944] discussed how to obtain more accurate measurements through experimental design. A two-pan balance holds an object of unknown mass in one pan and an object of known mass in the other pan, and when the plates level off, the masses are equal, and if two objects are weighed in opposite pans, then the difference in mass is obtained.

An investigator measures the mass of two apples A and B using a two-pan balance scale (Figure 2.21). Two apples are weighed in one pan of the scale, and then in opposite pans. Let σ^2 be the variance of an individual weighing. This scenario can be decsribed as two equations with the unknown weights w_1 and w_2 of the two apples.

$$w_1 + w_2 = m_1$$
$$w_1 - w_2 = m_2. \tag{2.8}$$

Solving (2.8) for w_1, w_2 leads to

$$\hat{w}_1 = (m_1 + m_2)/2$$
$$\hat{w}_2 = (m_1 - m_2)/2. \tag{2.9}$$

So, $Var(\hat{w}_1) = Var(\hat{w}_2) = \sigma^2/2$. This is half the value when the objects are weighed separately.

The moral of the story is that the data collection process has a significant impact on the precision of estimates.

This can also be viewed as a linear regression problem.

$$w_1 x_{11} + w_2 x_{21} = m_1$$
$$w_1 x_{21} - w_2 x_{21} = m_2,$$

(2.10)

where,

$$x_{ij} = \begin{cases} 1 & \text{if the } i^{th} \text{ measurement of the } j^{th} \text{ object is in the left pan} \\ -1 & \text{if the } i^{th} \text{ measurement of the } j^{th} \text{ object is in the right pan.} \end{cases}$$

(2.10) can be written in matrix form (2.4):

$$\begin{pmatrix} m_1 \\ m_2 \end{pmatrix} = \begin{pmatrix} 1 & 1 \\ 1 & -1 \end{pmatrix} \begin{pmatrix} w_1 \\ w_2 \end{pmatrix} + \begin{pmatrix} \epsilon_1 \\ \epsilon_2 \end{pmatrix}.$$

The least squares estimates are equal to (2.9).

FIGURE 2.21: Pan Balance

2.8.4   Computation Lab: Linear Regression

2.8.4.1   Childhood Mortality

The World Bank [worldbank.org, 2021] records under 5 childhood mortality rate on a global scale. Under5mort and LifeSatisfaction contain data on under 5 childhood mortality (per 1,000) and life satisfaction between 2005 and 2017 across an average of 225 countries or areas of the world in the data frame lifesat_childmort [ourworldindata.org, 2021].

The regression model in 2.8.1 can be estimated using the lm() function in R.

```
lifemortreg <- lm(log10(LifeSatisfaction) ~ log10(Under5mort),
                  data = lifesat_childmort)

broom::tidy(lifemortreg)
broom::glance(lifemortreg)
```

TABLE 2.7: Regression Estimates of Life Satisfaction vs. Under 5 Mortality

term	estimate	std.error	statistic	p.value
(Intercept)	0.8929	0.0044	204.00	0
log10(Under5mort)	-0.1333	0.0032	-41.28	0

TABLE 2.8: Regression Model Summary of Life Satisfaction vs. Under 5 Mortality

r.squared	adj.r.squared	sigma	statistic	p.value
0.5279	0.5276	0.0637	1704	0

broom::tidy(lifemortreg) provides a summary the regression model (see Table 2.7). Each row summarizes the estimates and tests for β_j: the first column term labels the regression coefficients; the second column estimate shows $\hat{\beta}_j$; the third column std.error shows $se(\beta_j)$, the fourth column statistic shows observed value of the t statistics for testing $H_0 : \beta_j = 0$; and the last column p.value shows the corresponding p-value of the tests. For example, $\hat{\beta}_0 = 0.8929$, and $\hat{\beta}_1 = 0.0044$.

Table 2.8 shows the first five columns of broom::glance(lifemortreg)—a summary of the entire regression model. For example, $R^2 = 0.5279$ and $\hat{\sigma}^2 = 0.0637$.

The assumptions of constant variance and normality of residuals are investigated in Figure 2.22 for model (2.2).

```
p1 <- lifemortreg %>%
  ggplot(aes(.fitted, .resid)) +
  geom_point() +
  geom_hline(yintercept = 0)

p2 <- lifemortreg %>%
  ggplot(aes(sample = .resid)) +
  geom_qq() +
  geom_qq_line()


p1 + p2
```

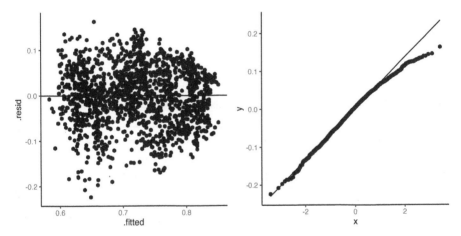

FIGURE 2.22: Regression Diagnostics for Life Satisfaction and Childhood Mortality

The left-hand panel in Figure 2.22 shows the residuals decreasing as the fitted values increase, indicating a non-constant variance; the right-hand panel in Figure 2.22 shows a normal Q-Q plot of the residuals that has a slightly heavier right-tail than the normal distribution. Heteroscedascity and lack of normality do not bias the least-squares estimates, but can increase the variances of the parameter estimates, so may affect R^2 and hypothesis tests of the parameters.

R can be used to fit a multiple regression model of `LifeSatisfaction` on `Region`. `lm()` converts `Region` from a character type into a factor type with seven levels, which is then converted into six dummy variables. The code below fits a linear regression model that has six dummy variables and an intercept term using `lm()`.

```
mod1 <- lm(LifeSatisfaction ~ Region,
            data = lifesat_childmort[lifesat_childmort$Year==2017,])
```

Table 2.9 shows the least least-squares estimates, where six estimates correspond to each region except East Asia & Pacific, which is used as the *reference category*. Each `Coefficient` $\hat{\beta}_k$ is an estimate of β_k in (2.7) for the kth region except the region used as the reference category (i.e., East Asia & Pacific). For example, $\hat{\beta}_0 = 5.5809$, $\hat{\beta}_1 = 0.51$, etc. How can we interpret these least squares estimates? Let's compute the mean of `LifeSatisfaction` by year. The least squares estimate is the mean difference between `Europe & Central Asia` and `East Asia & Pacific`, which is equal to $6.092 - 5.581 = 0.511$.

```
lifesat_childmort %>%
  filter(Year == 2017) %>%
  group_by(Region) %>%
  summarise(`Mean Life Satisfaction` =
            mean(LifeSatisfaction, na.rm = T))
```

```
## # A tibble: 7 x 2
##   Region                      `Mean Life Satisfaction`
##   <chr>                                          <dbl>
## 1 East Asia & Pacific                             5.58
## 2 Europe & Central Asia                           6.09
## 3 Latin America & Caribbean                       6.00
## 4 Middle East & North Africa                      5.35
## 5 North America                                   7.20
## 6 South Asia                                      4.32
## 7 Sub-Saharan Africa                              4.35
```

TABLE 2.9: Regression of Life Satisfaction and Region

term	estimate	std.error	statistic	p.value
(Intercept)	5.5809	0.2252	24.7831	0.0000
RegionEurope & Central Asia	0.5114	0.2580	1.9824	0.0494
RegionLatin America & Caribbean	0.4213	0.2948	1.4289	0.1553
RegionMiddle East & North Africa	-0.2311	0.3090	-0.7479	0.4558
RegionNorth America	1.6225	0.6565	2.4713	0.0147
RegionSouth Asia	-1.2615	0.4213	-2.9944	0.0033
RegionSub-Saharan Africa	-1.2276	0.2680	-4.5802	0.0000

2.8.4.2   Hotelling's Weighing Problem

We can find the least-squares estimates in Example 2.16 using R.

```
#step-by-step matrix multiplication for weighing problem
X <- matrix(c(1, 1, 1, -1), nrow = 2, ncol = 2) #define X matrix
Y <- t(X) %*% X #X'X
W <- solve(Y) #(X'X)^(-1)
```

```
W %*% t(X)  #(X'X)^(-1)*X'
```

```
      [,1] [,2]
[1,]  0.5  0.5
[2,]  0.5 -0.5
```

`W %*% t(X)` can be used in (2.5) to find the least-squares estimators of the apple weights: $(\hat{w}_1, \hat{w}_2)' = \begin{pmatrix} 0.5 & 0.5 \\ 0.5 & -0.5 \end{pmatrix} (m_1, m_2)'$.

```
W
```

```
      [,1] [,2]
[1,]  0.5  0.0
[2,]  0.0  0.5
```

W can be used in (2.6) to find the variance of the least-squares estimators

$$\text{cov}((\hat{w}_1, \hat{w}_2)') = \begin{pmatrix} 0.5 & 0.0 \\ 0.0 & 0.5 \end{pmatrix} \sigma^2.$$

It's also possible to use `lm()` to compute the least-squares estimates.

2.9   Exercises

Exercise 2.1. Figure 2.3 displays the frequency distribution from Example 2.5. Create a similar plot using `ggplot()` to display the *relative frequency* *distribution* for each treatment group instead of the frequency distribution. *(Hint: geom_bar(aes(y = ..prop.., group = trt))* constructs bars with the relative frequency distribution for each **trt** group.)

Exercise 2.2. Reproduce Figure 2.4 using `geom_histogram(aes(y = ..density..))` as shown below for Example 2.5. `ggplot_build()` extracts the computed values for the histogram. Use the extracted vales to confirm that the areas of all bars of the density histogram sum up to 1.

```
h <- covid19_trial %>%
  ggplot(aes(age)) +
  geom_histogram(
    aes(y = ..density..),
    binwidth = 2,
    color = 'black',
    fill = 'grey'
  )
densities <- ggplot_build(h)$data[[1]]
```

Exercise 2.3. In Section 2.3, 20 patients for Example 2.5 were randomly selected from a population of 25,000.

a. Explain why the probability of choosing a random sample of 20 from 25,000 is

$$\frac{1}{\binom{25000}{20}}.$$

b. Use R to randomly select 4 patients from a total of 8 patients, and calculate the probability of choosing this random sample.

Exercise 2.4. The treatment assignment displayed in Table 2.5 is from one treatment assignment out of $\binom{20}{10}$ possible random treatment assignments.

a. R can be used to generate all possible combinations of 10 patients assigned to the treatment group using `combn(covid_sample$id, 10)`. Use the command to generate the combinations and confirm that there are $\binom{20}{10}$ ways to generate the treatment assignments.

b. To generate the random array of treatment assignments in Table 2.5, R was used to randomly select 10 patients for the active treatment first and then assigned the placebo to the remaining 10. We can also generate a random treatment assignment vector by randomly selecting a column from the matrix of all possible treatment assignments created in part a. In R, randomly select a column and assign the first 10 to the active treatment and the rest to the placebo. Repeat a few times and discuss whether the procedure results in random assignments as desired. *(Hint: You can use* `if_else(condition, true, false)` *inside* `mutate` *to return* `true` *when* `condition` *is met and* `false` *otherwise.)*

c. Another scheme we may consider is to assign TRT or PLA to each patient with equal probabilities independently. Implement the procedure in R. Repeat a few times and discuss whether the procedure results in random assignments as desired.

d. Can you think of another procedure in R for randomly assigning 10 patients to active treatment? Implement your own procedure in R to verify the procedure results in random assignments.

Exercise 2.5. Show that

$$P(a < X < b) = F(b) - F(a)$$

for a continuous random variable X. Would this equality hold if X is a discrete random variable?

Exercise 2.6. In this exercise you will use R to compute the normal density and the cumulative distribution functions using `dnorm` and `pnorm`.

a. For any given values of `mu`, `sigma`, and `y`, what value does `dnorm(y, mu, sigma)` compute? Write the expression in terms of μ, σ, and y. How does the value compare to the value computed by `1/sigma*dnorm((y-mu)/sigma)`? Explain.

b. For any given values of `mu`, `sigma`, and `y`, what value does `pnorm(y, mu, sigma)` compute? Write the expression in terms of μ, σ, and y. How does the value compare to the value computed by `pnorm((y-mu)/sigma)`? Explain.

Exercise 2.7. Suppose $X \sim N(0, 1)$ and $W_n \sim \chi_n^2$ independently for any positive integer n. Let $V_n = X / \sqrt{W_n/n}$.

a. We know $V_n \sim t_n$. Show that V_n^2 follows an F distribution and specify the parameters.

b. Simulate 1,000 samples of V_1^2, V_8^2, and V_{50}^2 using the `rt` function in R and plot the histogram for each variable.

c. Identify density functions of V_1^2, V_8^2, and V_{50}^2 and use R to plot the theoretical density curves using `df()` from the histograms from part b. for each variable. Is the shape of the density the exact same as the histogram? Explain.

Exercise 2.8. Explain why the reference line in Figure 2.12 uses the first and third quartiles from the $Unif[0, 1]$, and ordered sample.

Exercise 2.9. Recall that simulated squared values from Exercise 2.7 follow F distributions. Use `geom_qq` and `geom_qq_line` to plot Q-Q plots of the simulated squared values against the appropriate F distributions. Comment on the plot.

Exercise 2.10. We generated Figure 2.16 by first computing the mean values for each column and then pivoting the values into a single column of mean values. Replicate the data generation steps and use the replicated data to reproduce the plot by first pivoting the data into the long format with a column for the simulation numbers and a column for the values and, then computing the mean values for each simulation using `group_by`. Verify that the new plot is identical to Figure 2.16.

Exercise 2.11. A chemist has seven light objects to weigh on a balance pan scale. The standard deviation of each weighing is denoted by σ. In a 1935 paper, Frank Yates [Yates, 1935] suggested an improved technique by weighing all seven objects together, and also weighing them in groups of three. The groups are chosen so that each object is weighed four times altogether, twice with any other object and twice without it.

Let $y_1, ..., y_8$ be the readings from the scale so that the equations for determining the unknown weights, $\beta_1, ..., \beta_7$, are

$$y_1 = \beta_1 + \beta_2 + \beta_3 + \beta_4 + \beta_5 + \beta_6 + \beta_7 + \epsilon_1$$
$$y_2 = \beta_1 + \beta_2 + \beta_3 + \epsilon_2$$
$$y_3 = \beta_1 + \beta_4 + \beta_5 + \epsilon_3$$
$$y_4 = \beta_1 + \beta_6 + \beta_7 + \epsilon_4$$
$$y_5 = \beta_2 + \beta_4 + \beta_6 + \epsilon_5$$
$$y_6 = \beta_2 + \beta_5 + \beta_7 + \epsilon_6$$
$$y_7 = \beta_3 + \beta_4 + \beta_7 + \epsilon_7$$
$$y_8 = \beta_3 + \beta_5 + \beta_6 + \epsilon_8,$$

where the $\epsilon_i, i = 1, ..., 8$ are independent errors.

Hotelling [1944] suggested modifying Yates' procedure by placing in the other pan of the scale those of the objects not included in one of his weighings. In other words if the first three objects are to be weighed, then the remaining four objects would be placed in the opposite pan.

a. Write Yates' procedure in matrix form and find the least squares estimates of β.

b. Write Hotelling's procedure in matrix form $\mathbf{y} = X\beta + \epsilon$, where $\mathbf{y}' = (y_1, ..., y_8)$, $\beta' = (\beta_1, ..., \beta_7)$, $\epsilon' = (\epsilon_1, ..., \epsilon_8)$, and X is an 8×7 matrix. Find the least squares estimate of β.

c. Find the variance of a weight using Yates' and Hotelling's procedures.

d. If the chemist wanted estimates of the weights with the highest precision, then which procedure (Yates or Hotelling) would you recommend that the chemist use to weigh objects? Explain your reasoning.

Exercise 2.12. Does life satisfaction change by region over time? Use the `lifesat_childmort` data from Example 2.15 to explore this question.

3

Comparing Two Treatments

3.1 Introduction

Consider the following scenario. Volunteers for a medical study are randomly assigned to two groups to investigate which group has a higher mortality rate. One group receives the standard treatment for the disease, and the other group receives an experimental treatment. Since people were randomly assigned to the two groups the two groups of patients should be similar *except for the treatment they received.*

If the group receiving the experimental treatment lives longer on average and the difference in survival is both practically meaningful and statistically significant then **because of the randomized design** it's reasonable to infer that the new treatment *caused* patients to live longer. Randomization is supposed to ensure that the groups will be similar with respect to both *measured* and *unmeasured* factors that affect study participants' mortality.

Consider two treatments labelled A and B. In other words interest lies in a single factor with two levels. Examples of study objectives that lead to comparing two treatments are:

- Is fertilizer A or B better for growing wheat?
- Is a new vaccine compared to placebo effective at preventing COVID-19 infections?
- Will web page design A or B lead to different sales volumes?

These are all examples of comparing two *treatments*. In experimental design, *treatments* are different procedures applied to *experimental units*—the plots, patients, web pages to which we apply *treatments*.

In the first example, the treatments are two fertilizers and the experimental units might be plots of land. In the second example, the treatments are an active vaccine and **placebo** vaccine (a sham vaccine) to prevent COVID-19, and the experimental units are volunteers that consented to participate in a vaccine study. In the third example, the treatments are two web page designs and the website visitors are the experimental units.

DOI: 10.1201/9781003033691-3

3.2　Treatment Assignment Mechanism and Propensity Score

In a randomized experiment, the treatment assignment mechanism is developed and controlled by the investigator, and the probability of an assignment of treatments to the units is known *before* data is collected. Conversely, in a non-randomized experiment, the assignment mechanism and probability of treatment assignments are unknown to the investigator.

Suppose, for example, that an investigator wishes to randomly assign two experimental units, unit 1 and unit 2, to two treatments (A and B). Table 3.1 shows all possible treatment assignments.

TABLE 3.1: All Possible Treatment Assignments: Two Units, Two Treatments

Treatment Assignment	unit1	unit2
1	A	A
2	B	A
3	A	B
4	B	B

3.2.1　Propensity Score

The probability that an experimental unit receives treatment is called the **propensity score**. In this case, the probability that an experimental unit receives treatment A (or B) is $1/2$.

It's important to note that the probability of a treatment assignment and propensity scores are different probabilities, although in some designs they may be equal.

In general, if there are N experimental units and two treatments then there are 2^N possible treatment assignments.

3.2.2　Assignment Mechanism

There are four possible treatment assignments when there are two experimental units and two treatments. The probability of a particular treatment assignment is $1/4$. This probability is called the **assignment mechanism**. It is the probability that a particular treatment assignment will occur (see Section 7.2 for further discussion).

3.2.3 Computation Lab: Treatment Assignment Mechanism and Propensity Score

`expand.grid()` was used to compute Table 3.1. This function takes the possible treatments for each unit and returns a data frame containing one row for each combination. Each row corresponds to a possible **randomization** or **treatment assignment**.

```
expand.grid(unit1 = c("A","B"), unit2 = c("A","B"))
```

3.3 Completely Randomized Designs

In the case where there are two units and two treatments, it wouldn't be a very informative experiment if both units received A or both received B. Therefore, it makes sense to rule out this scenario. If we rule out this scenario then we want to assign treatments to units such that one unit receives A and the other receives B. There are two possible treatment assignments: treatment assignments 2 and 3 in Table 3.1. The probability of a treatment assignment is $1/2$, and the probability that an individual patient receives treatment A (or B) is still $1/2$.

A **completely randomized experiment** has the number of units assigned to treatment A, N_A, fixed in advance so that the number of units assigned to treatment B, $N_B = N - N_A$, is also fixed in advance. In such a design, N_A units are randomly selected, from a population of N units, to receive treatment A, with the remaining N_B units assigned to treatment B. Each unit has probability N_A/N of being assigned to treatment A.

How many ways can N_A experimental units be selected from N experimental units such that the order of selection doesn't matter and replacement is not allowed (i.e., a unit cannot be selected more than once)? This is the same as the distinct number of treatment assignments. There are $\binom{N}{N_A}$ distinct treatment assignments with N_A units out of N assigned to treatment A. Therefore, the assignment mechanism or the probability of any particular treatment assignment is $1/\binom{N}{N_A}$.

Example 3.1 (Comparing Fertilizers). Is fertilizer A better than fertilizer B for growing wheat? It is decided to take one large plot of land and divide it into twelve smaller plots of land, then treat some plots with fertilizer A, and some with fertilizer B. How should we assign fertilizers (*treatments*) to plots of land (Table 3.2)?

Some of the plots get more sunlight and not all the plots have the exact same soil composition which may affect wheat yield. In other words, the plots are not identical. Nevertheless, we want to make sure that we can identify the treatment effect even though the plots are not identical. Statisticians sometimes state this as being able to identify the treatment effect (*viz.* difference between fertilizers) in the presence of other sources of variation (*viz.* differences between plots).

Ideally, we would assign fertilizer A to six plots and fertilizer B to six plots. How can this be done so that the only differences between plots is fertilizer type? One way to assign the two fertilizers to the plots is to use six playing cards labelled A (for fertilizer A) and six playing cards labelled B (for fertilizer B), shuffle the cards, and then assign the first card to plot 1, the second card to plot 2, etc.

TABLE 3.2: Observed Treatment Assignment in Example 3.1

Plot 1, B, 11.4	Plot 4, A, 16.5	Plot 7, A, 26.9	Plot 10, B, 28.5
Plot 2, A, 23.7	Plot 5, A, 21.1	Plot 8, B, 26.6	Plot 11, B, 14.2
Plot 3, B, 17.9	Plot 6, A, 19.6	Plot 9, A, 25.3	Plot 12, B, 24.3

3.3.1 Computation Lab: Completely Randomized Experiments

How can R be used to assign treatments to plots in Example 3.1? Create **cards** as a vector of 6 A's and 6 B's, and use the **sample()** function to generate a random permutation (i.e., shuffle) of **cards**.

```
cards <- c(rep("A",6),rep("B",6))
set.seed(1)
shuffle <- sample(cards)
shuffle
```

```
 [1] "B" "A" "B" "A" "A" "A" "A" "B" "A" "B" "B" "B"
```

This can be used to assign B to the first plot, and A to the second plot, etc. The full treatment assignment is shown in Table 3.2.

3.4 The Randomization Distribution

The treatment assignment in Example 3.1 is the one that the investigator used to collect the data in Table 3.2. This is one of the $\binom{12}{6} = 924$ possible ways of allocating 6 A's and 6 B's to the 12 plots. The probability of choosing any of these treatment allocations is $1/\binom{12}{6} = 0.001$.

TABLE 3.3: Mean and Standard Deviation of Fertilizer Yield in Example 3.1

Treatment	Mean yield	Standard deviation yield
A	22.18	3.858
B	20.48	6.999

The mean and standard deviation of the outcome variable, yield, under treatment A is $\bar{y}_A^{obs} = 22.18$, $s_A^{obs} = 3.86$, and under treatment B is $\bar{y}_B^{obs} = 20.48$, $s_B^{obs} = 7$. The observed difference in mean yield is $\hat{\delta}^{obs} = \bar{y}_A^{obs} - \bar{y}_B^{obs} = 1.7$ (see Table 3.3). The superscript *obs* refers to the statistic calculated under the treatment assignment used to collect the data or the *observed* treatment assignment.

The distributions of a sample can also be described by the *empirical cumulative distribution function (ECDF)* (see Figure 3.1):

$$\hat{F}(y) = \frac{\sum_{i=1}^{n} I(y_i \le y)}{n},$$

where n is the number of sample points and $I(\cdot)$ is the indicator function

$$I(y_i \le y) = \begin{cases} 1 & \text{if } y_i \le y \\ 0 & \text{if } y_i > y \end{cases}$$

TABLE 3.4: Random Shuffle of Treatment Assignment in Example 3.1

Plot 1, A, 11.4	Plot 4, B, 16.5	Plot 7, B, 26.9	Plot 10, B, 28.5
Plot 2, A, 23.7	Plot 5, A, 21.1	Plot 8, A, 26.6	Plot 11, A, 14.2
Plot 3, B, 17.9	Plot 6, B, 19.6	Plot 9, B, 25.3	Plot 12, A, 24.3

Is the difference in wheat yield due to the fertilizers or chance?

- Assume that there is no difference in the average yield between fertilizer A and fertilizer B.

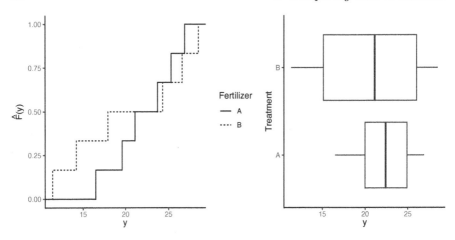

FIGURE 3.1: Distribution of Yield

- If there is no difference then the yield would be the same even if a different treatment allocation occurred.

- Under this assumption of no difference between the treatments, if one of the other 924 treatment allocations (e.g., A, A, B, B, A, B, B, A, B, B, A, A) was used then the treatments assigned to plots would have been **randomly shuffled**, but the yield in each plot would be exactly the same as in Table 3.2. This *shuffled* treatment allocation is shown in Table 3.4, and the difference in mean yield for this allocation is δ = -2.23 (recall that the observed treatment difference $\hat{\delta}^{obs}$ = 1.7).

A probability distribution for $\delta = \bar{y}_A - \bar{y}_B$, called the **randomization distribution**, is constructed by calculating δ for each possible randomization (i.e., treatment allocation).

Investigators are interested in determining whether fertilizer A produces a higher yield compared to fertilizer B, which corresponds to null and alternative hypotheses

H_0 : Fertilizers A and B have the same mean wheat yield,

H_1 : Fertilizer B has a greater mean wheat yield than fertilizer A.

3.4.1　Computation Lab: Randomization Distribution

The data from Example 3.1 is in the `fertdat` data frame. The code chunk below computes the randomization distribution.

```
# 1. Number of possible randomizations

N <- choose(12,6)

delta <- numeric(N) # Store the results

# 2. Generate N treatment assignments

trt_assignments <- combn(1:12,6)

for (i in 1:N){
  # 3. compute mean difference and store the result

  delta[i] <- mean(fertdat$fert[trt_assignments[,i]]) -
    mean(fertdat$fert[-trt_assignments[,i]])
}
```

1. N is the total number of possible treatment assignments or randomizations.

2. `trt_assignments <- combn(1:12,6)` generates all combinations of 6 elements taken from `1:12` (i.e., 1 through 12) as a 6×924 matrix, where the ith column `trt_assignments[,i]`, $i = 1, \ldots, 924$, represents the experimental units assigned to treatment A.

3. `fertdat$fert[trt_assignments[,i]]` selects `fertdat$fert` values indexed by `trt_assignments[,i]`. These values are assigned to treatment A. `fertdat$fert[-trt_assignments[,i]]` drops `fertdat$fert` values indexed by `trt_assignments[,i]`. These values are assigned to treatment B.

3.5 The Randomization p-value

3.5.1 One-sided p-value

Let T be a test statistic such as the difference between treatment means or medians. The p-value of the *randomization test* $H_0 : T = 0$ can be calculated as the probability of obtaining a test statistic as extreme or more extreme than the observed value of the test statistic t^* (i.e., in favour of H_1). The p-value is the proportion of randomizations as extreme or more extreme than the observed value of the test statistic t^*.

Definition 3.1 (One-sided Randomization p-value). Let T be a test statistic and t^* the observed value of T. The one-sided p-value to test $H_0 : T = 0$ is defined as:

$$P(T \geq t^*) = \sum_{i=1}^{\binom{N}{N_A}} \frac{I(t_i \geq t^*)}{\binom{N}{N_A}}, \text{ if } H_1 : T > 0;$$

$$P(T \leq t^*) = \sum_{i=1}^{\binom{N}{N_A}} \frac{I(t_i \leq t^*)}{\binom{N}{N_A}}, \text{ if } H_1 : T < 0.$$

A hypothesis test to answer the question posed in Example 3.1 is $H_0 : \delta = 0$ v.s. $H_1 : \delta > 0$, $\delta = \bar{y}_A - \bar{y}_B$. The observed value of the test statistic is 1.7.

3.5.2   Two-sided Randomization p-value

If we are using a two-sided alternative, then how do we calculate the randomization p-value? The randomization distribution may not be symmetric, so there is no justification for simply doubling the probability in one tail.

Definition 3.2 (Two-sided Randomization p-value). Let T be a test statistic and t^* the observed value of T. The two-sided p-value to test $H_0 : T = 0$ vs. $H_1 : T \neq 0$ is defined as:

$$P(|T| \geq |t^*|) = \sum_{i=1}^{\binom{N}{N_A}} \frac{I(|t_i| \geq |t^*|)}{\binom{N}{N_A}}.$$

The numerator counts the number of randomizations where either t_i or $-t_i$ exceed $|t^*|$.

3.5.3   Computation Lab: Randomization p-value

The randomization distribution was computed in Section 3.4.1, and stored in `delta`. We want to compute the proportion of randomizations that exceed `obs_diff`.

```
glimpse(delta)
```

```
## num [1:924] -5.93 -3.5 -3.6 -4.03 -2.97 ...
```

`delta >= obs_diff` creates a Boolean vector that is `TRUE` if `delta >=` `obs_diff`, and `FALSE` otherwise, and sum applied to this Boolean vector counts the number of `TRUE`.

```
obs_diff
```

```
## [1] 1.7
```

```
N
```

```
## [1] 924
```

```
pval <- sum(delta >= obs_diff)/N
pval
```

```
## [1] 0.303
```

The p-value can be interpreted as the proportion of randomizations that would produce an observed mean difference between A and B of at least 1.7 assuming the null hypothesis is true. In other words, under the assumption that there is no difference between the treatment means, 30.3% of randomizations would produce as extreme or more extreme difference than the observed mean difference of 1.7.

The two-sided p-value to test if there is a difference between fertilizers A and B in Example 3.1 can be computed as

```
pval2side <- sum(delta >= abs(obs_diff) |
                  (-1)*delta >= abs(obs_diff))/N
pval2side
```

```
## [1] 0.6061
```

In this case, the randomization distribution is roughly symmetric, so the two-sided p-value is approximately double the one-sided p-value.

The R code to produce Figure 3.2, without annotations, is shown below. The plot displays the randomization distribution of $\delta = \bar{y}_A - \bar{y}_B$ for Example 3.1. The left panel shows the distribution using $1 - \hat{F}_\delta$, and the dotted line indicates how to read the p-value from this graph, and the right panel shows a histogram where the black bars show the values more extreme than the observed value.

```
p1 <- tibble(delta) %>%
  ggplot(aes(delta)) +
  geom_histogram(bins = 30, colour = "white",
                 aes(y = ..count.. / sum(..count..))) +
  scale_x_continuous(breaks = seq(-10, 10, by = 2))

p2 <- tibble(delta) %>%
  ggplot(aes(delta)) +
  geom_step(aes(y = 1 - ..y..), stat = "ecdf") +
  scale_y_continuous(breaks = seq(0, 1, by = 0.1)) +
  scale_x_continuous(breaks = seq(-10, 10, by = 2))

p2 + p1
```

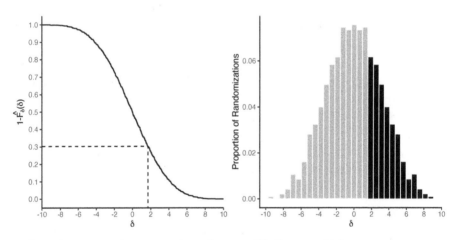

FIGURE 3.2: Randomization Distribution of Difference of Means

3.5.4 Randomization Confidence Intervals

Consider a completely randomized design comparing two groups where the treatment effect is additive. In Example 3.1, suppose that the yields for fertilizer A were shifted by Δ, these shifted responses; $y_{i_A} - \Delta$ should be similar to y_{i_B} for $i = 1, \ldots, 6$, and the randomization test on these two sets of responses should not reject H_0. In other words the difference between the distribution of yield for fertilizers A and B can be removed by subtracting Δ from each plot assigned to fertilizer A.

Loosely speaking, a confidence interval, for the mean difference, consists of all the plausible values of the parameter Δ. A **randomization confidence interval** can be constructed by considering all values of Δ_0 for which the randomization test does not reject $H_0 : \Delta = \Delta_0$ vs. $H_a : \Delta \neq \Delta_0$.

Definition 3.3 (Randomization Confidence Interval). Let T_Δ be the test statistic calculated using the treatment responses for treatment A shifted by Δ, t^* its observed value, and $p(\Delta) = F_{T_\Delta}(t^*_\Delta) = P(T_\Delta \leq t^*_\Delta)$ be the observed value of the CDF as a function of Δ.

A $100(1 - \alpha)\%$ **randomization confidence interval** for Δ can then be obtained by inverting $p(\Delta)$. A two-sided $100(1 - \alpha)\%$ is (Δ_L, Δ_U), where $\Delta_L = p^{-1}(\alpha/2) = \max_{p(\Delta \leq \alpha/2)} \Delta$, and $\Delta_U = p^{-1}(1 - \alpha/2) = \min_{p(\Delta \leq 1-\alpha/2)} \Delta$ [Ernst, 2004].

3.5.5 Computation Lab: Randomization Confidence Intervals

Computing Δ_L, Δ_U involves recomputing the randomization distribution of T_Δ for a series of values $\Delta_1, \ldots, \Delta_k$. This can be done by trial and error, or by a search method (see for example Garthwaite [1996]).

In this section, a trial and error method is implemented using a series of R functions.

The function `randomization_dist()` computes the randomization distribution for the mean difference in a randomized two-sample design.

```
randomization_dist <- function(dat, M, m) {
  N <- choose(M, m)
  randdist <- numeric(N)
  trt_assignments <- combn(1:M, m)
  for (i in 1:N) {
    randdist[i] <-
      mean(dat[trt_assignments[, i]]) -
      mean(dat[-trt_assignments[, i]])
  }
  return(randdist)
}
```

The function `randomization_pctiles()` computes $p(\Delta)$ for a sequence of trial values for Δ.

```
randomization_pctiles <- function(delta, datA, datB, M, m) {
  pctiles <- numeric(length(delta))
  for (i in 1:length(delta)) {
    yA <- datA + delta[i]
    yB <- datB
    obs_diff <- mean(yA) - mean(yB)
    dat <- c(yA, yB)
    y <- randomization_dist(dat, M, m)
    pctiles[i] <- sum(y <= obs_diff) / N
  }
  return(pctiles)
}
```

The function `randomization_ci()` computes the Δ_L, Δ_U as well as the confidence level of the interval.

```
randomization_ci <- function(alpha, delta, yB, yA, M, m) {
  ptiles <- randomization_pctiles(delta, yB, yA, M, m)
  L <- sum(ptiles <= alpha / 2)
  U <- sum(ptiles <= (1 - (alpha / 2)))
  Lptile <- ptiles[L]
  Uptile <- ptiles[U]
  LCI <- delta[L]
  UCI <- delta[U]
  conf_level <- Lptile + 1 - Uptile
  return(data.frame(Lptile, Uptile, conf_level, LCI, UCI))
}
```

Example 3.2 (Confidence interval for wheat yield in Example 3.1). A 99% randomization confidence interval for the wheat data can be obtained by using `randomization_ci()`. The data for the two groups are defined by `yA` and `yB`, the confidence level is `alpha`, with M total experimental units and m experimental units in one of the groups. The sequence of values for Δ is found by trial and error, but it's important that the tails of the distribution of Δ are computed far enough so that we have values for the upper and lower $\alpha/2$ percentiles.

A plot of $p(\Delta)$ is shown in Figure 3.3. `delta` is selected so that `pdelta` is computed in tails of the distribution of T_Δ.

```
alpha <- 0.01
delta <- seq(-16,16,by=0.1)
M <- 12
m <- 6
yA <- fertdat$fert[fertdat$shuffle=="A"]
yB <- fertdat$fert[fertdat$shuffle=="B"]

tibble(pdelta = randomization_pctiles(delta = delta,
                                      datA = yA,
                                      datB = yB,
                                      M = M,
                                      m = m)) %>%
  ggplot(aes(delta, pdelta)) + geom_point()
```

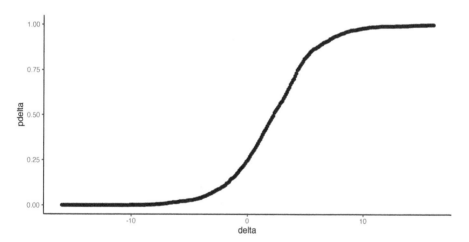

FIGURE 3.3: Distribution of Δ in Example 3.2

```
randomization_ci(alpha, delta, yA,yB, M, m)
```

```
##     Lptile Uptile conf_level LCI UCI
## 1 0.004329 0.9946    0.00974  -8  14
```

`Lptile` and `Uptile` are the lower and upper percentiles of the distribution of T_Δ used for the confidence interval, `conf_level` is actual confidence level of the confidence interval, and finally `LCI`, `UCI` are the limits of a $(1-\texttt{conf\_level})$ level confidence interval. In this case, $(\Delta_L, \Delta_U) = (-8, 14)$ is a 99.03% confidence interval for the difference between the means of treatments A and B.

3.6   Randomization Distribution of a Test Statistic

Test statistics other than $T = \bar{y}_A - \bar{y}_B$ could be used to measure the effectiveness of fertilizer A in Example 3.1. Investigators may wish to compare differences between medians, standard deviations, odds ratios, or other test statistics.

3.6.1   Computation Lab: Randomization Distribution of a Test Statistic

The randomization distribution of the difference in group medians can be obtained by modifying the `randomization_dist()` function (see 3.5.5) used to calculate the difference in group means. We can add `func` as an argument to `randomization_dist()` and modify the function so that the type of difference can be specified.

```
randomization_dist <- function(func, dat, M, m) {
  N <- choose(M, m)
  randdist <- numeric(N)
  trt_assignments <- combn(1:M, m)
  for (i in 1:N) {
    randdist[i] <-
      (func(dat[trt_assignments[, i]]) -
         func(dat[-trt_assignments[, i]]))
  }
  return(randdist)
}
```

The randomization distribution of the difference in medians is

```
randdist_median <- randomization_dist(median, fertdat$fert, 12, 6)
```

The p-value of the randomization test comparing two medians is

```
obs_median_diff <- median(yA) - median(yB)
pval <- sum(randdist_median >= obs_median_diff)/N
pval
```

```
[1] 0.5195
```

3.7   Computing the Randomization Distribution using Monte Carlo Sampling

Computation of the randomization distribution involves calculating the test statistic for every possible way to split the data into two samples of size N_A. If $N = 100$ and $N_A = 50$, this would result in $\binom{100}{50} = 1.0089 \times 10^{20}$ billion differences. These types of calculations are not practical unless the sample size is small.

Instead, we can resort to Monte Carlo sampling from the randomization distribution to estimate the exact p-value.

The data set can be randomly divided into two groups and the test statistic calculated. Several thousand test statistics are usually sufficient to get an accurate estimate of the exact p-value and sampling can be done without replacement.

If M test statistics, t_i, $i = 1, ..., M$ are randomly sampled from the permutation distribution, a one-sided Monte Carlo p-value for a test of $H_0 : \mu_T = 0$ versus $H_1 : \mu_T > 0$ is

$$\hat{p} = \frac{1 + \sum_{i=1}^{M} I(t_i \geq t^*)}{M + 1}.$$

Including the observed value t^* there are $M + 1$ test statistics.

3.7.1   Computation Lab: Calculating the Randomization Distribution using Monte Carlo Sampling

Example 3.3 (What is the effect of caffeine on reaction time?). There is scientific evidence that caffeine reduces reaction time [Santos et al., 2014, McLellan et al. [2016]]. A study of the effects of caffeine on reaction time was conducted on a group of 100 high school students. The investigators randomly assigned an equal number of students to two groups: one group (CAFF) consumed a caffeinated beverage prior to taking the test, and the other group (NOCAFF) consumed the same amount of water. The research objective was to study the effect of caffeine on reaction time to test the hypothesis that caffeine would reduce reaction time among high school students. The data from the study is in the data frame `rtdat`.

```
ggplot(rtdat, aes(rt, group)) + geom_boxplot()
```

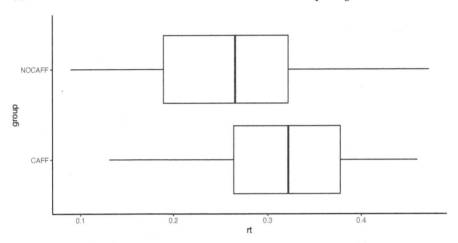

The data indicate that the difference in median reaction times between the CAFF and NOCAFF groups is 0.056 seconds. Is the observed difference due to random chance or is there evidence it is due to caffeine? Let's try to calculate the randomization distribution using `randomization_dist()`.

```
randomization_dist(func = mean, dat = rt, M = 100, m = 50)
```

```
Error in numeric(N): vector size specified is too large
```

Currently, R can only support vectors up to 2^{52} elements [Wickham, 2019], so computing the full randomization distribution becomes much more difficult. In this case, Monte Carlo sampling provides a feasible way to approximate the randomization distribution and p-value.

```
caff <- rtdat$rt[rtdat$group == "CAFF"]
ncaff <- rtdat$rt[rtdat$group == "NOCAFF"]
rt1 <- c(caff, ncaff)
N <- 10000
result <- numeric(N)

for (i in 1:N) {
  index <- sample(length(rt),
                  size = length(caff),
                  replace = FALSE)
  result[i] <- median(rt[index]) - median(rt[-index])
}
```

```
observed <- median(caff) - median(ncaff)
phatval <- (sum(result >= observed) + 1) / (N + 1)
phatval
```

```
## [1] 0.0038
```

A p-value equal to 0.004 indicates that the median difference is unusual assuming the null hypothesis is true. Thus, this study provides evidence that caffeine slows down reaction time.

3.8 Properties of the Randomization Test

The p-value of the randomization test must be a multiple of $1/\binom{N}{N_A}$. If a significance level of $\alpha = k/\binom{N}{N_A}$, where $k = 1, ..., \binom{N}{N_A}$ is chosen, then $P(\text{type I}) = \alpha$. In other words, the randomization test is an exact test.

If α is not chosen as a multiple of $1/\binom{N}{N_A}$, but $k/\binom{N}{N_A}$ is the largest p-value less than α, then $P(\text{type I}) = k/\binom{N}{N_A} < \alpha$, and the randomization test is conservative. Either way, the test is guaranteed to control the probability of a type I error under very minimal conditions: randomization of the experimental units to the treatments [Ernst, 2004].

3.9 The Two-sample t-test

Consider designing a study where the primary objective is to compare a continuous variable in each group. Let Y_{ik} be the observed outcome for the ith experimental unit in the kth treatment group, for $i = 1, ..., n_k$ and $k = 1, 2$. The outcomes in the two groups are assumed to be independent and normally distributed with different means but an equal variance σ^2, $Y_{ik} \sim N(\mu_k, \sigma^2)$.

Let $\theta = \mu_1 - \mu_2$, be the difference in means between the two treatments. $H_0 : \theta = \theta_0$ vs. $H_1 : \theta \neq \theta_0$ specify a hypothesis test to evaluate whether the evidence shows that the two treatments are different.

The sample mean for each group is given by $\bar{Y}_k = (1/n_k)\sum_{i=1}^{n_k} Y_{ik}$, $k = 1, 2$, and the pooled sample variance is

$$S_p^2 = \frac{(n_1 - 1)S_1^2 + (n_2 - 1)S_2^2}{(n_1 + n_2 - 2)},$$

where S_k^2 is the sample variance for group $k = 1, 2$.

The two-sample t statistic is given by

$$T = \frac{\bar{Y}_1 - \bar{Y}_2 - \theta_0}{S_p\sqrt{(1/n_1 + 1/n_2)}}. \tag{3.1}$$

When H_0 is true, $T \sim t_{n_1+n_2-2}$.

For example, the two-sided p-value for testing θ is $P\left(|t_{n_1+n_2-2}| > |T^{obs}|\right)$, where T^{obs} is the observed value of (3.1). The hypothesis testing procedure assesses the strength of evidence contained in the data against the null hypothesis. If the p-value is adequately small, say, less than 0.05 under a two-sided test, we reject the null hypothesis and claim that there is a significant difference between the two treatments; otherwise, there is no significant difference and the study is inconclusive.

In Example, 3.1 $H_0 : \mu_A = \mu_B$ and $H_1 : \mu_A < \mu_B$. The pooled sample variance and the observed value of the two-sample t-statistic for this example are:

$$S_p^2 = \frac{(n_1 - 1)S_1^2 + (n_2 - 1)S_2^2}{n_1 + n_2 - 2} = 5.65,$$

and

$$T^{obs} = \frac{\bar{y}_A - \bar{y}_b}{S_p\sqrt{(1/n_A + 1/n_B)}} = \frac{20.22 - 22.45}{5.65\sqrt{(1/6 + 1/6)}} = -0.69.$$

The p-value is $P(t_{10} < -0.69) = 0.3$. There is little evidence that fertilizer A produces higher yields than B.

3.9.1   Computation Lab: Two-sample t-test

We can use R to compute the p-value of the two-sample t-test for Example 3.1. Recall that the data frame `fertdat` contains the data for this example: `fert` is the yield and `shuffle` is the treatment.

```
glimpse(fertdat)
```

```
## Rows: 12
## Columns: 2
## $ fert    <dbl> 11.4, 23.7, 17.9, 16.5, 21.1, 19.6, 2~
## $ shuffle <chr> "A", "A", "B", "B", "A", "B", "B", "A~
```

The pooled variance s_p^2 and observed value of the two-sample t statistic are:

```
yA <- fertdat$fert[fertdat$shuffle == "A"]
yB <- fertdat$fert[fertdat$shuffle == "B"]

n1 <- length(yA) - 1
n2 <- length(yB) - 1

sp <- sqrt(((n1 - 1)*var(yA) + (n2 - 1)*var(yB))/(n1 + n2 -2))

sp
```

```
## [1] 5.595
```

```
tobs <- (mean(yA)-mean(yB))/(sp*sqrt(1/6 + 1/6))
tobs
```

```
## [1] -0.6914
```

The observed value of the two-sample t-statistic is -0.6914.

Finally, the p-value for this test can be calculated using the CDF of the t_n, where $n = 6 + 6 - 2 = 10$.

```
df <- 6 + 6 - 2
pt(tobs, df)
```

```
## [1] 0.2525
```

These calculations are also implemented in `stats::t.test()`.

```
t.test(yA, yB, var.equal = TRUE, alternative = "less")
```

```
##
##  Two Sample t-test
##
## data:  yA and yB
```

```
## t = -0.69, df = 10, p-value = 0.3
## alternative hypothesis: true difference in means is less than 0
## 95 percent confidence interval:
##    -Inf 3.621
## sample estimates:
## mean of x mean of y
##      20.22      22.45
```

The assumption of normality can be checked using normal quantile plots, although the t-test is robust against non-normality.

```
fertdat %>%
  ggplot(aes(sample = fert, linetype = shuffle)) +
  geom_qq(aes(shape = shuffle)) +
  geom_qq_line() +
  ggtitle("Normal Quantile Plot")
```

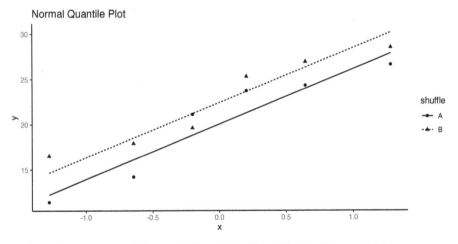

FIGURE 3.4: Normal Quantile Plot of Fertilizer Yield in Example 3.1

Figure 3.4 indicates that the normality assumption is satisfied, although the sample sizes are fairly small.

Notice that the p-value from the randomization test and the p-value from two-sample t-test are almost identical although the randomization test neither depends on normality nor independence. The randomization test does depend on Fisher's concept that after randomization, if the null hypothesis is true, the two results obtained from each particular plot will be *exchangeable*. The randomization test tells you what you could say if exchangeability were true.

3.10   Blocking

Randomizing subjects to two treatments **should** produce two treatment groups where all the covariates are balanced (i.e., have similar distributions), but it doesn't **guarantee** that the treatment groups will be balanced on all covariates. In many applications there may be covariates that are *known a priori* to have an effect on the outcome, and it's important that these covariates be measured and balanced, so that the treatment comparison is not affected by the imbalance in these covariates. Suppose an important covariate is income level (low/medium/high). If income level is related to the outcome of interest, then it's important that the two treatment groups have a balanced number of subjects in each income level, and this shouldn't be left to chance. To avoid an imbalance between income levels in the two treatment groups, the design can be *blocked* by income group, by separately randomizing subjects in low, medium, and high income groups.

Example 3.4. Kim et al. [2017] conducted a randomized clinical trial to evaluate hemoglobin level (an important component of blood) levels after a surgery to remove a cancer. Patients were randomized to receieve a new treatment or placebo. The study was conducted at seven major institutions in the Republic of Korea. Previous research has shown that the amount of cancer in a person's body, measured by cancer stage (stage I—less cancer, stages II—IV - more cancer), has an effect on hemoglobin. 450 (225 per group) patients were required to detect a significant difference in the main study outcome at the 5% level (with 90% power - see Chapter—4).

To illustrate the importance of blocking, consider a realistic, although hypothetical, scenario related to Example 3.4. Suppose that among patients eligible for inclusion in the study, 1/3 have stage I cancer, and 225 (50%) patients are randomized to the treatment and placebo groups. Table 3.5 shows that the distribution of Stage in the placebo group is different than the distribution in the Treatment group. In other words, the distribution of cancer stage in each treatment group is unbalanced. The imbalance in cancer stage might create a bias when comparing the two treatment groups since it's known a priori that cancer stage has an effect on the main study outcome (hemoglobin level after surgery). An unbiased comparison of the treatment groups would have Stage balanced between the two groups.

How can an investigator guarantee that Stage is balanced in the two groups? Separate randomizations by cancer stage, *blocking* by cancer stage.

TABLE 3.5: Distribution of Cancer Stage by Treatment Group in Example 3.4 using Unrestricted Randomization

Stage	Placebo	Treatment
Stage I	70	80
Stage II-IV	155	145

TABLE 3.6: Distribution of Cancer Stage by Treatment Group in Example 3.4 using Restricted Randomization

Stage	Placebo	Treatment
Stage I	75	75
Stage II-IV	150	150

3.11 Treatment Assignment in Randomized Block Designs

If Stage was balanced in the two treatment groups in Example 3.4, then 50% of stage I patients would receive Placebo, and 50% Treatment. If we *block* or *separate* the randomizations by Stage, then this will yield treatment groups balanced by stage. There will be $\binom{150}{75}$ randomizations for the stage I patients, and $\binom{300}{150}$ randomizations for the stage II-IV patients. Table 3.6 shows the results of block randomization.

3.11.1 Computation Lab: Generating a Randomized Block Design

Let's return to Example 3.4, and suppose that we are designing a study where 450 subjects will be randomized to two treatments, and 1/3 of the 450 subjects (150) have stage I cancer. `cancerdat` is a data frame containing a patient `id` and `Stage` information.

```
cancerdat_stageI <- data.frame(id = 1:150)
```

First, we will create a data frame, `cancerdat_stageI` of patient `id` with stage I cancers. Next, randomly select 50% of patient `id` in this block `sample(cancerdat_stageI$id, floor(nrow(cancerdat_stageI)/2))` and

assign these patients to `Treatment`, and the remaining to `Placebo` using
`treat = ifelse(id %in% trtids, "Treatment", "Placebo")`.

```
trtids <-
  sample(cancerdat_stageI$id, floor(nrow(cancerdat_stageI) / 2))

cancerdat_Ibl <-
  mutate(cancerdat_stageI,
         treat = ifelse(id %in% trtids, "Treatment", "Placebo"))
```

The treatment assignments for the first 4 patients with stage I cancer are
shown below.

```
cancerdat_Ibl %>% head(n = 4)
```

```
##   id  Stage       treat
## 1  1 Stage I    Placebo
## 2  2 Stage I    Placebo
## 3  3 Stage I  Treatment
## 4  4 Stage I  Treatment
```

3.12 Randomized Matched Pairs Design

A randomized matched pairs design arranges experimental units in pairs, and
treatment is randomized within a pair. In other words, each experimental unit
is a block. In Chapter 5, we will see how this idea can be extended to compare
more than two treatments using randomized block designs.

Example 3.5 (Wear of boys' shoes). Measurements on the amount of wear
of the soles of shoes worn by 10 boys were obtained by the following design
(this example is based on 3.2 in Box et al. [2005]).

Each boy wore a special pair of shoes with the soles made of two different
synthetic materials, A (a standard material) and B (a cheaper material). The
left or right sole was randomly assigned to A or B, and the amount of wear
after one week was recorded (a smaller value means less wear). During the test
some boys scuffed their shoes more than others, but each boy's shoes were
subjected to the same amount of wear.

In this case, each boy is a block, and the two treatments are randomized within a block.

Material was randomized to the left or right shoe by flipping a coin. The observed treatment assignment is one of $2^{10} = 1,024$ equiprobable treatment assignments.

The observed mean difference is 1.7. Figure 3.5, a connected dot plot of wear for each boy, shows material B had higher wear for most boys.

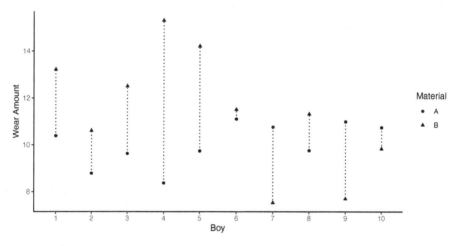

FIGURE 3.5: Boy's Shoe Example

3.13 Randomized Matched Pairs versus Completely Randomized Design

Ultimately, the goal is to compare units that are similar except for the treatment they were assigned. So, if groups of similar units can be created before randomizing, then it's reasonable to expect that there should be less variability between the treatment groups. Blocking factors are used when the investigator has knowledge of a factor **before** the study that is measurable and might be strongly associated with the dependent variable.

The most basic blocked design is the randomized pairs design. This design has n units where two treatments are randomly assigned to each unit which results in a pair of observations (X_i, Y_i), $i = 1, \ldots, n$ on each unit. In this case, each unit is a block. Assume that the $X's$ and $Y's$ have means μ_X and μ_Y,

TABLE 3.7: Possible Randomizations for Example 3.5

Observed	L	L	R	R	R	L	R	R	L	L
Possible	R	R	R	R	L	R	R	R	L	L
Wear (A)	10.39	8.79	9.64	8.37	9.74	11.1	10.76	9.76	10.99	10.74
Wear (B)	13.22	10.61	12.51	15.31	14.21	11.51	7.54	11.31	7.7	9.84

and variances σ_X^2 and σ_Y^2, and the pairs are independently distributed and $Cov(X_i, Y_i) = \sigma_{XY}$. An estimate of $\mu_X - \mu_Y$ is $\bar{D} = \bar{X} - \bar{Y}$. It follows that

$$E\left(\bar{D}\right) = \mu_X - \mu_Y$$
$$Var\left(\bar{D}\right) = \frac{1}{n}\left(\sigma_X^2 + \sigma_Y^2 - 2\rho\sigma_X\sigma_Y\right),$$

(3.2)

where ρ is the correlation between X and Y.

Alternatively, if n units had been assigned to two independent treatment groups (i.e., $2n$ units) then $Var\left(\bar{D}\right) = (1/n)\left(\sigma_X^2 + \sigma_Y^2\right)$. Comparing the variances we see that the variance of $\bar{D}$ is smaller in the paired design if the correlation is positive. So, pairing is a more effective experimental design.

3.14 The Randomization Test for a Randomized Paires Design

Table 3.7 shows the observed (randomization) and another possible (randomization) for material A in Example 3.5. If the other possible randomization was observed, then $\bar{y}_A - \bar{y}_B = -1.4$.

The differences $\bar{y}_A - \bar{y}_B$ can be analyzed so that we have one response per boy. Under the null hypothesis, the wear of a boy's left or right shoe is the same regardless of what material he had on his sole, and the material assigned is based on the result of, for example, a sequence of ten tosses of a fair coin (e.g., in R this could be implemented by `sample(x = c("L","R"),size = 10,replace = TRUE)`). This means that under the null hypothesis if the Possible Randomization in Table 3.7 was observed, then for the first boy the right side would have been assigned material A and the left side material B, but the amount of wear on the left and right shoes would be the same, so the difference for the first boy would have been 2.8 instead of -2.8 since his wear for materials A and B would have been 13.22 and 10.39 respectively.

The randomization distribution is obtained by calculating 1,024 averages $\bar{y}_A - \bar{y}_B = (\pm -2.8 \pm -1.8 \pm \cdots \pm 0.9)/ 10$, corresponding to each of the $2^{10} = 1,024$ possible treatment assignments.

3.14.1   Computation Lab: Randomization Test for a Paired Design

The data for Example 3.5 is in **shoedat_obs** data frame.

```
head(shoedat_obs , n = 3)
```

```
##    boy sideA sideB  wearA wearB
## 1    1    L     R  10.390 13.22
## 2    2    L     R   8.792 10.61
## 3    3    R     L   9.636 12.51
```

The code chunk below generates the randomization distribution.

```
diff <- shoedat_obs %>%
  mutate(diff = wearA - wearB) %>%
  select(diff) %>%
  unlist() # calculate difference

N <- 2 ^ (10) # number of treatment assignments

res <- numeric(N) #vector to store results

LR <- list(c(-1, 1)) # difference is multiplied by -1 or 1

# generate all possible treatment assign
trtassign <- expand.grid(rep(LR, 10))

for (i in 1:N) {
  res[i] <- mean(as.numeric(trtassign[i, ]) * diff)
}
```

The 2^{10} treatment assignments are computed using **expand.grid()** on a list of 10 vectors (**c(-1,1)**)—each element of the list is the *potential* sign of the difference for one experimental unit (i.e., boy), and **expand.grid()** creates a data frame from all combinations of these 10 vectors.

```
tibble(res) %>%
  ggplot(aes(res)) +
  geom_histogram(bins=20, colour = "black", fill = "grey") +
  xlab("Mean Difference") +
  geom_vline(aes(xintercept = obs_diff), size =1.5)
```

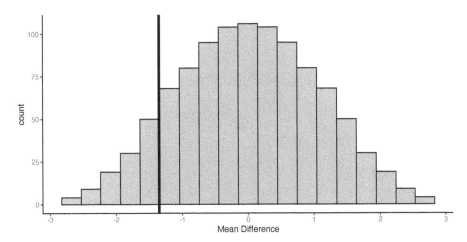

FIGURE 3.6: Randomization Distribution–Boys' Shoes

The p-value for testing if B has more wear than A is

$$P(D \leq d^*) = \sum_{i=1}^{2^{10}} \frac{I(d_i \leq d^*)}{2^{10}},$$

where $D = \bar{y}_A - \bar{y}_B$, and d^* is the observed mean difference.

```
sum(res <= obs_diff) / N # p-value
```

```
## [1] 0.1084
```

The value of $d^* = -1.3$ is not unusual under the null hypothesis since only
111 (i.e., 10%) differences of the randomization distribution are less than -1.3.
Therefore, there is no evidence of a significant increase in the amount of wear
with the cheaper material B.

3.15   Paired t-test

If we assume that the differences from Example 3.5 are a random sample from a normal distribution, then $t = \sqrt{10}\bar{d}/S_{\bar{d}} \sim t_{10-1}$, where, $S_{\bar{d}}$ is the sample standard deviation of the paired differences. The p-value for testing if $\bar{D} < 0$ is $P(t_9 < t)$. In other words, this is the same as a one-sample t-test of the differences.

3.15.1   Computation Lab: Paired t-test

In Section 3.14.1, diff is a vector of the differences for each boy in Example 3.5. The observed value of the t-statistic for the one-sample test can be computed.

```
n <- length(diff)
t_pair_obs <- sqrt(n) * mean(diff) / sd(diff)
t_pair_obs
```

```
## [1] -1.319
```

The p-value for testing $H_0 : \bar{D} = 0$ versus $H_a : \bar{D} < 0$ is

```
df <- n - 1
pt(t_pair_obs, df)
```

```
## [1] 0.1099
```

Alternatively, t.test() can be used.

```
t.test(diff, alternative = "less")
```

```
##
##   One Sample t-test
##
## data:  diff
## t = -1.3, df = 9, p-value = 0.1
## alternative hypothesis: true mean is less than 0
```

```
## 95 percent confidence interval:
##     -Inf 0.5257
## sample estimates:
## mean of x
##     -1.348
```

3.16 Exercises

Exercise 3.1. Suppose $X_1 \sim N(10, 25)$ and $X_2 \sim N(5, 4)$ in a population. You randomly select 100 samples from the population and assign treatment A to half of the sample and B to the rest. Simulate the sample with treatment assignments and the covariates, X_1 and X_2. Compare the distributions of X_1 and X_2 in the two treatment groups. Repeat the simulation one hundred times. Do you observe consistent results?

Exercise 3.2. Identify *treatments* and *experimental units* in the following scenarios.

a. City A would like to evaluate whether a new employment training program for the unemployed is more effective compared to the existing program. The City decides to run a pilot program for selected employment program applicants.

b. Marketing Agency B creates and places targeted advertisements on video-sharing platforms for its clients. The Agency decides to run an experiment to compare the effectiveness of placing advertisements before vs. during vs. after videos.

c. Paul enjoys baking and finds a new recipe for chocolate chip cookies. Paul decides to test it by bringing cookies baked using his current recipe and the new recipe to his study group. Each member of the group blindly tastes each kind and provides their ratings.

Exercise 3.3. A study has three experimental units and two treatments—A and B. List all possible treatment assignments for the study. How many are there? In general, show that there are 2^N possible treatment assignments for an experiement with N experimenal units and 2 treatments.

Exercise 3.4. Consider the scenario in Example 3.1, and suppose that an investigator only has enough fertilizer A to use on four plots. Answer the following questions.

a. What is the probability that an individual plot receives fertilizer A?

b. What is the probability of choosing the treatment assignment A, A, A, A, B, B, B, B, B, B, B, B?

Exercise 3.5. Show that the one-sided p-value is $1 - \hat{F}_T\left(t^*\right)$ if $H_1 : T > 0$ and $\hat{F}_T\left(t^*\right)$ if $H_1 : T < 0$, where $\hat{F}_T$ is the ECDF of the randomization distribution of T and t^* is the observed value of T.

Exercise 3.6. Show that the two-sided p-value is $1 - \hat{F}_T\left(|t^*|\right) + \hat{F}_T\left(-|t^*|\right)$, where $\hat{F}_T$ is the ECDF of the randomization distribution of T and t^* is the observed value of T.

Exercise 3.7. The actual confidence level `conf_level` does not equal the theoretical confidence level 0.01 in Example 3.2. Explain why.

Exercise 3.8. Consider Example 3.5. For each of the 10 boys, we randomly assigned the left or right sole to material A and the remaining side to B. Use R's `sample` function to simulate a treatment assignment.

Exercise 3.9. Recall that the randomization test for the data in Example 3.5 fails to find evidence of a significant increase in the amount of wear with material B. Does this mean that material B has equivalent wear to material A? Explain.

Exercise 3.10. Consider the study from Example 3.4. Recall that the clinical trial consists of 450 patients. 150 of the patients have stage I cancer and the rest have stages II-IV cancer. In Computation Lab: Generating a Randomized Block Design, we created a balanced treatment assignment for the stage I cancer patients.

a. Create a balanced treatment assignment for the stage II-IV cancer patients.

b. Combine treatment assignments for stage I and stage II-IV. Show that the distribution of stage is balanced in the overall treatment assignment.

Exercise 3.11. Consider a randomized pair design with n units where two treatments are randomly assigned to each unit, resulting in a pair of observations (X_i, Y_i), for $i = 1, \ldots, n$ on each unit. Assume that $E[X_i] = \mu_X$, $E[Y_i] = \mu_y$, and $Var(X_i) = Var(Y_i) = \sigma^2$ for $i = 1, \ldots, n$. Alternatively, we may consider an unpaired design where we assign two independent treatment groups to $2n$ units.

a. Show that the ratio of the variances in the paired to the unpaired design is $1 - \rho$, where ρ is the correlation between X_i and Y_i.

b. If $\rho = 0.5$, how many subjects are required in the unpaired design to yield the same precision as the paired design?

Exercise 3.12. Suppose that two drugs A and B are to be tested on 12 subjects' eyes. The drugs will be randomly assigned to the left eye or right eye based on the flip of a fair coin. If the coin toss is heads then a subject will receive drug A in their right eye. The coin was flipped 12 times and the following sequence of heads and tails was obtained:

$$T \quad T \quad H \quad T \quad H \quad T \quad T \quad T \quad H \quad T \quad T \quad H$$

a. Create a table that shows how the treatments will be allocated to the 12 subjects' left and right eyes.

b. What is the probability of obtaining this treatment allocation?

c. What type of experimental design has been used to assign treatments to subjects? Explain.

4

Power and Sample Size

4.1 Introduction

Consider a scenario where a continuous study outcome, with variance equal to one, is measured in two groups. The group means are μ_1 and μ_2. An investigator tests $H_0 : \mu_1 - \mu_2 = 0$ versus $H_1 : \mu_1 \neq \mu_2$ with a sample of fifteen experimental units in each group.

The function my_ttest() simulates data to test this hypothesis with a t-test, where $\mu_1 = 0$, $\mu_2 = 0.1$, and $\sigma^2 = 1$ in both groups.

```
my_ttest <- function(sampsize) {
  set.seed(6)
  x1 <- rnorm(sampsize, mean = 0, sd = 1)
  x2 <- rnorm(sampsize, mean = 0.1, sd = 1)
  myttest <- t.test(x1, x2, var.equal = TRUE)
  return(myttest$p.value)
  }
```

The simulation below shows that the t-test failed to detect a significant difference even though a difference exists (i.e., H_0 is false).

```
my_ttest(15)
```

```
## [1] 0.4447
```

In this case, we know $\mu_1 \neq \mu_2$, so why does the the p-value indicate that there is no difference between the group means? Let's change the sample size to 1,000 per group.

```
my_ttest(1000)
```

DOI: 10.1201/9781003033691-4

[1] 0.004706

Now, the p-value indicates that there is strong evidence that H_0 is false. The test failed to detect a difference when the sample size was 15 due to low **power**, and in this case, low power is due to a small sample size in each group.

Power is important in many fields of study. Suppose that researchers would like to compare two versions of a web page to investigate whether one page leads to an increase in sales. A pharmaceutical company is comparing a novel treatment for cancer against the standard treatment to investigate whether patient mortality decreases when receiving the novel treatment. A psychologist is studying how physical expression influences psychological processes such as risk taking, so she plans to randomize subjects into two groups where one group was instructed to pose in a high-power (this has nothing to do with statistical power) position and the second group in a low-power position. In all these scenarios, the investigators can calculate how many experimental units are required so that the statistical test used will have a high probability of detecting a significant difference between the groups if indeed there is a difference.

In studies with large sample sizes in each group the power might be large, but the differences detected may be small and not practically important.

4.2 Statistical Hypotheses and the Number of Experimental Units

Suppose that experimental units are randomized to treatments A or B with equal probability. Let μ_A and μ_B be the mean responses in groups A and B. The null hypothesis is that there is no difference between A and B; the alternative claims there is a difference: $H_0 : \mu_A = \mu_B$ vs $H_1 : \mu_A \neq \mu_B$

The type I error rate is defined as:

$$\alpha = P\left(\text{type I error}\right) = P_{H_0}\left(\text{Reject } H_0\right).$$

P_{H_0} means the probability calculated using the distribution induced by H_0. For example, a testing of μ, the meanof a normal distribution, $H_0 : \mu = \mu_0$ vs. $H_1 : \mu \neq \mu_0$, at the 5% level would have p-value $= P_{H_0}\left(\text{Reject } H_0\right) = P\left(|t_{n-1}| > |t^{obs}|\right)$, with $t^{obs} = \sqrt{n}(\bar{x} - \mu_0)/S_x$.

The type II error rate is defined as:

$$\beta = P\left(\text{type II error}\right) = P_{H_1}\left(\text{Accept } H_0\right).$$

Power is defined as:

$$\text{power} = 1 - \beta$$
$$= 1 - P_{H_1}\left(\text{Accept } H_0\right) = P_{H_1}\left(\text{Reject } H_0\right).$$

4.3   Power of the One-Sample z-test

Let $X_1, ..., X_n$ be a random sample from a $N(\mu, \sigma^2)$ distribution. A test of the hypothesis

$$H_0 : \mu = \mu_0 \text{ versus } H_1 : \mu \neq \mu_0$$

will reject at level α if and only if

$$\left| \frac{\bar{X} - \mu_0}{\sigma/\sqrt{n}} \right| \geq z_{1-\alpha/2},$$

or

$$\left| \bar{X} - \mu_0 \right| \geq \frac{\sigma}{\sqrt{n}} z_{1-\alpha/2},$$

where z_p is the pth quantile of the $N(0, 1)$.

The power of the test at $\mu = \mu_1$ (i.e., when $H_1 : \mu = \mu_1$) is

$$1 - \beta = 1 - P\left(\text{type II error}\right)$$
$$= P_{H_1}\left(\text{Reject } H_0\right)$$
$$= P_{H_1}\left(\left| \bar{X} - \mu_0 \right| \geq \frac{\sigma}{\sqrt{n}} z_{1-\alpha/2}\right)$$
$$= P_{H_1}\left(\bar{X} - \mu_0 \geq \frac{\sigma}{\sqrt{n}} z_{1-\alpha/2}\right) + P_{H_1}\left(\bar{X} - \mu_0 < \frac{-\sigma}{\sqrt{n}} z_{1-\alpha/2}\right).$$

Subtract the mean μ_1 and divide by $\sigma/\sqrt{n}$ to obtain:

$$1 - \beta = 1 - \Phi\left(z_{1-\alpha/2} - \left(\frac{\mu_1 - \mu_0}{\sigma/\sqrt{n}}\right)\right) + \Phi\left(-z_{1-\alpha/2} - \left(\frac{\mu_1 - \mu_0}{\sigma/\sqrt{n}}\right)\right).$$

The power function of the one-sample z-test depends on α, μ_1, μ_0, σ, and n.

4.3.1   Computation Lab: Power of the One-Sample z-test

An R function to compute the power of the one-sample z-test (4.3) using
`qnorm()` to calculate $z_{1-\alpha/2}$ and `pnorm()` to calculate $\Phi(\cdot)$ is

```
pow_ztest <- function(alpha, mu1, mu0, sigma, n) {
    arg1 <- qnorm(1 - alpha / 2) - (mu1 - mu0) / (sigma / sqrt(n))
    arg2 <- -1 * qnorm(1 - alpha / 2) -
    (mu1 - mu0) / (sigma / sqrt(n))
    return(1 - pnorm(arg1) + pnorm(arg2))
}
```

Example 4.1. Calculate the power of $H_0 : \mu = 0$ vs. $H_1 : \mu = 0.2$ with 30
experimental units per group, $\sigma = 0.2, \alpha = 0.05$.

```
pow_ztest(alpha = 0.05, mu1 = 0.15, mu0 = 0, sigma = 0.2, n = 30)
```

```
## [1] 0.9841
```

A study with 30 experimental units will have a 0.98 probability of detecting a
mean difference of 0.2, when $\sigma = 0.2, \alpha = 0.05$.

4.4   Power of the One-Sample t-test

Let $X_1, ..., X_n$ be i.i.d. $N(\mu, \sigma^2)$. A test of the hypothesis

$$H_0 : \mu = \mu_0 \text{ versus } H_1 : \mu \neq \mu_0$$

will reject at level α if and only if

$$\left| \frac{\bar{X} - \mu_0}{S/\sqrt{n}} \right| \geq t_{n-1, 1-\alpha/2},$$

where $t_{n-1,p}$ is the pth quantile of the t_{n-1}.

It can be shown that

$$\sqrt{n} \left[\frac{\bar{X} - \mu_0}{S} \right] = \frac{Z + \gamma}{\sqrt{V/(n-1)}},$$

where,

$$Z = \frac{\sqrt{n}(\bar{X} - \mu_1)}{\sigma},$$

$$\gamma = \frac{\sqrt{n}(\mu_1 - \mu_0)}{\sigma}, \text{ and}$$

$$V = \frac{(n-1)}{\sigma^2}S^2.$$

$Z \sim N(0,1)$, $V \sim \chi^2_{n-1}$, and Z is independent of V.

If $\gamma = 0$ then $\sqrt{n}\left[\frac{\bar{X} - \mu_0}{S}\right] \sim t_{n-1}$. But, if $\gamma \neq 0$, then $\sqrt{n}\left[\frac{\bar{X} - \mu_0}{S}\right] \sim t_{n-1,\gamma}$, where $t_{n-1,\gamma}$ is the non-central t-distribution with non-centrality parameter γ. If $\gamma = 0$, this is sometimes called the central t-distribution (see Figure 4.1).

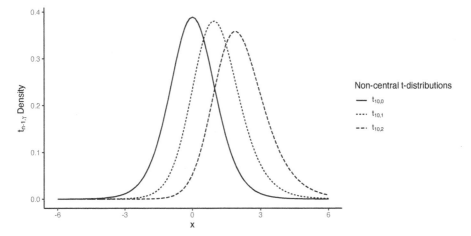

FIGURE 4.1: Noncentral t-distribution

The power of the test at $\mu = \mu_1$ is

$$
\begin{aligned}
1 - \beta &= 1 - P\,(\text{type II error}) \\
&= P_{H_1}\,(\text{Reject } H_0) \\
&= P_{H_1}\,(\text{Reject } H_0) \\
&= P_{H_1}\left(\left|\frac{\bar{X} - \mu_0}{\frac{S}{\sqrt{n}}}\right| \geq t_{n-1,1-\alpha/2}\right) \\
&= P_{H_1}\left(\frac{\bar{X} - \mu_0}{\frac{S}{\sqrt{n}}} \geq t_{n-1,1-\alpha/2}\right) + P_{H_1}\left(\frac{\bar{X} - \mu_0}{\frac{S}{\sqrt{n}}} < -t_{n-1,1-\alpha/2}\right) \\
&= P(t_{n-1,\gamma} \geq t_{n-1,1-\alpha/2}) + P(t_{n-1,\gamma} < -t_{n-1,1-\alpha/2}).
\end{aligned}
$$

4.4.1   Computation Lab: Power of the One-Sample t-test

The following function calculates the power function using (4.4) for the one-sample t-test in R:

```
onesampttestpow <- function(alpha, n, mu0, mu1, sigma) {
    delta <- mu1 - mu0
    t.crit <- qt(1 - alpha / 2, n - 1)
    t.gamma <- sqrt(n) * (delta / sigma)
    t.power <-
        1 - pt(t.crit, n - 1, ncp = t.gamma) +
        pt(-t.crit, n - 1, ncp = t.gamma)
    return(t.power)
}
```

Example 4.2. Calculate the power of $H_0 : \mu = 0$ vs. $H_1 : \mu = 0.15$ with $n = 10$, $\sigma = 0.2$, and $\alpha = 0.05$ by calling the above function.

```
onesampttestpow(
    alpha = 0.05,
    n = 10,
    mu0 = 0,
    mu1 = 0.15,
    sigma = 0.2
)
```

```
## [1] 0.562
```

This means that the test will reject H_0 in 56.2% of studies testing this hypothesis (i.e., the study testing this hypothesis was replicated a large number of times).

`stats::power.t.test()` is part of the default R packages. Using this function on the previous example we get

```
power.t.test(
    n = 10,
    delta = 0.15,
    sd = 0.2,
    sig.level = 0.05,
    type = "one.sample"
)
```

```
##
##          One-sample t test power calculation
##
##                   n = 10
##               delta = 0.15
##                  sd = 0.2
##           sig.level = 0.05
##               power = 0.5619
##         alternative = two.sided
```

Exactly one of the parameters n, delta, power, sd, and sig.level must be passed as NULL, and that parameter is determined from the others. In this example, the function calculates power given the other parameters. If sample size is required for, say, 80% power then use

```
power.t.test(
  power = 0.8,
  delta = 0.15,
  sd = 0.2,
  sig.level = 0.05,
  type = "one.sample"
)
```

```
##
##          One-sample t test power calculation
##
##                   n = 15.98
##               delta = 0.15
##                  sd = 0.2
##           sig.level = 0.05
##               power = 0.8
##         alternative = two.sided
```

The calculation shows that a study testing $H_0 : \mu = \mu_0$ vs $H_1 : \mu = \mu_1$, where delta= $\mu_0 - \mu_1$ requires sixteen experimental units for 80% power.

4.5 Power of the Two-Sample t-test

The two-sample t-test, described in Section 3.9, is often used to test if the difference between two means is zero. This section develops the power of this test using the same notation as Section 3.9.

Recall from Section 3.9 that T_n is the two-sample t-statistic, where $T_n \sim t_{n_1+n_2-2}$ under H_0 and $t_{n_1+n_2-2,\gamma}$ with non-centrality parameter $\gamma = \theta/\sigma\sqrt{1/n_1 + 1/n_2}$, under H_1.

H_0 is rejected if $|T_n| \geq t_{n_1+n_2-2,1-\alpha/2}$ where $t_{\lambda,p}$ is the *pth* quantile of the central t-distribution with λ degrees of freedom.

The sample size can be determined by specifying the type I and type II error rates, the standard deviation, and the difference in treatment means that the study aims to detect.

In a derivation similar to the power of the one-sample t-test in Section 4.4, the power of the two-sample t-test is

$$1 - \beta = P\left(t_{n_1+n_2-2,\gamma} \geq t_{n_1+n_2-2,\gamma,1-\alpha/2}\right) + P\left(t_{n_1+n_2-2,\gamma} < -t_{n_1+n_2-2,\gamma,1-\alpha/2}\right),$$

where $t_{\lambda,\gamma,p}$ is the *pth* quantile of the non-central t-distribution with non-centrality parameter γ and λ degrees of freedom. The power is not a closed form expression so it's not possible to derive a formula for the study sample size. Nevertheless, if the variance is assumed to be known, and if $n_1 = n_2 = n/2$, then the total sample size for a study at level α, power $1 - \beta$, that tests $H_0 : \theta = \theta_0$ vs. $H_1 : \theta = \theta_1$ is

$$n = \frac{4\sigma^2 \left(z_{1-\alpha/2} + z_{1-\beta}\right)}{\theta_1^2}. \tag{4.1}$$

4.5.1 Effect Size

In some studies, instead of specifying the difference in treatment means and standard deviation separately, the ratio

$$\text{ES} = \frac{\mu_1 - \mu_2}{\sigma}$$

can be specified. This ratio is called the scaled effect size. Cohen [1992] suggests that effect sizes of 0.2, 0.5, and 0.8 correspond to small, medium, and large effects.

4.5.2 Sample Size—Known Variance and Equal Allocation

Consider a study where experimental units are randomized into two treatment groups and the investigator would like an equal number of experimental units in each group.

If the variance is known, then a test statistic for testing if the means of two populations are equal is

$$Z = \frac{\bar{Y}_1 - \bar{Y}_2}{\sigma\sqrt{(1/n_1 + 1/n_2)}} \sim N(0, 1).$$

This is known as the two-sample z-test.

The power at $\theta = \theta_1$ is given by

$$1 - \beta = P\left(Z \geq z_{1-\alpha/2} - \frac{\theta_1}{\sigma\sqrt{1/n_1 + 1/n_2}}\right) + P\left(Z < -z_{1-\alpha/2} - \frac{\theta_1}{\sigma\sqrt{1/n_1 + 1/n_2}}\right).$$
(4.2)

Ignoring terms smaller than $\alpha/2$ and combining positive and negative θ in (4.2)

$$\beta \approx \Phi\left(z_{1-\alpha/2} - \frac{|\theta_1|}{\sigma\sqrt{1/n_1 + 1/n_2}}\right).$$
(4.3)

Applying Φ^{-1} to (4.3) and using 2.1, it follows that

$$z_{1-\beta} + z_{1-\alpha/2} = \left(\frac{|\theta_1|}{\sigma\sqrt{1/n_1 + 1/n_2}}\right).$$
(4.4)

If we assume that there will be an equal allocation of subjects to each group, then $n_1 = n_2 = n/2$, and the total sample size is

$$n = \frac{4\sigma^2\left(z_{1-\beta} + z_{1-\alpha/2}\right)^2}{\theta^2},$$
(4.5)

where $\alpha, \beta \in (0, 1)$.

4.5.3 Sample Size—Known Variance and Unequal Allocation

In many studies comparing two treatments, it is desirable to put more experimental units into the experimental group to learn more about this treatment. If the allocation of experimental units between the two groups is $r = n_1/n_2$ then $n_1 = r \cdot n_2$. Plugging this into (4.4) the sample size for n_2 is

$$n_2 = \frac{(1 + 1/r)\sigma^2\left(z_\beta + z_{\alpha/2}\right)^2}{\theta^2}.$$
(4.6)

4.5.4 Computation Lab: Power of the Two-Sample t-test

The following R function uses qt() to compute $t_{n,\gamma,\lambda}$ and pt() to compute the $t_{n,\gamma}$ CDF.

```
twosampttestpow <- function(alpha, n1, n2, mu1, mu2, sigma) {
  delta <- mu1 - mu2
  t.crit <- qt(1 - alpha / 2, n1 + n2 - 2)
  t.gamma <- delta / (sigma * sqrt(1 / n1 + 1 / n2))
  t.power <-
    (1 - pt(t.crit, n1 + n2 - 2, ncp = t.gamma) +
        pt(-t.crit, n1 + n2 - 2, ncp = t.gamma))
  return(t.power)
}
```

The power of a study to detect $\theta = 1$ with $\sigma = 3, n_1 = n_2 = 50$ is

```
twosampttestpow(
  alpha = .05,
  n1 = 50,
  n2 = 50,
  mu1 = 1,
  mu2 = 2,
  sigma = 3
)
```

```
## [1] 0.3786
```

stats::power.t.test() can also be used and gives the same results.

```
power.t.test(
  n = 50,
  delta = 1,
  sd = 3,
  sig.level = 0.05
)
```

```
##
##        Two-sample t test power calculation
```

```
##
##                    n = 50
##                delta = 1
##                   sd = 3
##            sig.level = 0.05
##                power = 0.3784
##          alternative = two.sided
##
## NOTE: n is number in *each* group
```

So the study would require 50 subjects per group to achieve 38% power to detect a difference of $\theta = 1$ at the 5% significance level assuming $\sigma = 3$. power.t.test() can also output the number of subjects required to achieve a certain power. Suppose the investigators want to know how many subjects per group would have to be enrolled in each group to achieve 80% power under the same conditions?

```
power.t.test(
  power = 0.8,
  delta = 1,
  sd = 3,
  sig.level = 0.05
)
```

```
##
##        Two-sample t test power calculation
##
##                    n = 142.2
##                delta = 1
##                   sd = 3
##            sig.level = 0.05
##                power = 0.8
##          alternative = two.sided
##
## NOTE: n is number in *each* group
```

142 subjects would be required in each group to achieve 80% power.

Figure 4.2 shows power of the two-sample t-test as a function of n and θ (left), and n and σ (right) with a horizontal line drawn at 0.8 power for reference.

```r
f <- function(n, d, sd)
  power.t.test(
    n = n,
    delta = d,
    sd = sd,
    type = "two.sample",
    alternative = "two.sided",
    sig.level = 0.05
  )$power

p1 <-
  ggplot() +
  xlim(0, 3.0) +
  scale_y_continuous(breaks = seq(0, 1, by = 0.1)) +
  geom_function(fun = f,
                args = list(n = 30, sd = 1.5),
                aes(linetype = "N = 30")) +
  geom_function(fun = f,
                args = list(n = 20, sd = 1.5),
                aes(linetype = "N = 20")) +
  geom_function(fun = f,
                args = list(n = 10, sd = 1.5),
                aes(linetype = "N = 10")) +
  ylab("Power") +
  xlab(TeX("$\\theta$")) +
  geom_hline(yintercept = 0.8) +
  guides(linetype = guide_legend(title = "Sample size"))

p2 <-
  ggplot() +
  xlim(0.15, 3.0) +
  scale_y_continuous(breaks = seq(0, 1, by = 0.1)) +
  geom_function(fun = f,
                args = list(n = 30, d = 1),
                aes(linetype = "N = 30")) +
  geom_function(fun = f,
                args = list(n = 20, d = 1),
                aes(linetype = "N = 20")) +
  geom_function(fun = f,
                args = list(n = 10, d = 1),
                aes(linetype = "N = 10")) +
  ylab("Power") +
  xlab(TeX("$\\sigma$")) +
  geom_hline(yintercept = 0.8) +
```

```
guides(linetype = guide_legend(title = "Sample size"))

p1 + p2
```

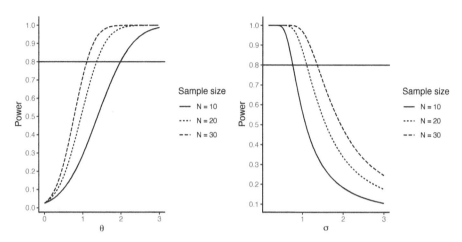

FIGURE 4.2: Power of Two-Sample t-test

Power as a function of effect size can be investigated by rearranging (4.2). In the R function below we assume $n_1 = n_2 = 10,]\alpha = 0.05$.

```
pow.t <- function(ES) {
  alpha <- 0.05
  nA <- 10
  nB <- 10
  t.crit <- qt(1 - alpha / 2, nA + nB - 2)
  t.gamma <- ES / (sqrt(1 / nA + 1 / nB))
  t.power <- (1 - pt(t.crit, nA + nB - 2, ncp = t.gamma) +
                pt(-t.crit, nA + nB - 2, ncp = t.gamma))
  t.power
}
```

The code below was used to create Figure 4.3.

```
ggplot() +
  scale_x_continuous(limits = c(-2, 2),
                     breaks = seq(-2, 2, by = 0.5)) +
```

```
scale_y_continuous(limits = c(0, 1),
                   breaks = seq(0, 1, by = 0.1)) +
geom_function(fun = pow.t) +
ylab("Power") +
xlab("Effect Size")
```

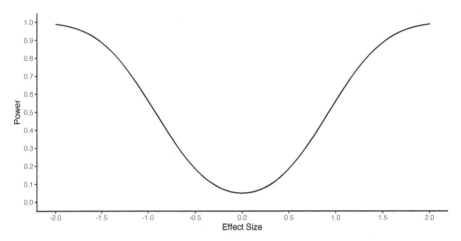

FIGURE 4.3: Two-Sample t-test Power and Effect Size, N = 10

The R code below defines a function to compute the sample size in groups 1 and 2 for unequal allocation using (4.6).

```
size2z.uneq.test <- function(theta, alpha, beta, sigma, r)
{
  zalpha <- qnorm(1 - alpha / 2)
  zbeta <- qnorm(1 - beta)
  n2 <- (1 + 1 / r) * (sigma * (zalpha + zbeta) / theta) ^ 2
  n1 <- r * n2
  c(n1, n2)
}
```

The number of patients in group 1 is computed using **size2z.uneq.test**.

```
# sample size for theta =1, alpha = 0.05,
# beta = 0.1, sigma = 2, r = 2
```

```
size2z.uneq.test(
  theta = 1,
  alpha = .05,
  beta = .1,
  sigma = 2,
  r = 2
)[1] # group 1 sample size (experimental group)
```

```
## [1] 126.1
```

So, the number of patients in the experimental group is 126. The number in the control group is computed below.

```
size2z.uneq.test(
  theta = 1,
  alpha = .05,
  beta = .1,
  sigma = 2,
  r = 2
)[2] # group 2 sample size (control group)
```

```
## [1] 63.04
```

The sample size required for 90% power to detect $\theta = 1$ with $\sigma = 2$ at the 5% level in a trial where two patients will be enrolled in the experimental arm for every patient enrolled in the control arm is 126 in the control group and 63 in the experimental group. The total sample size is 189.

The power of the two-sample z-test (4.3) can be studied as a function of the allocation ratio r. The code chunk below produces Figure 4.4.

```
# power of z test as a function of allocation ratio r,
# total sample size n, alpha, theta, and sigma
pow_ztest <- function(r, n, alpha, theta, sigma)
{
  n2 <- n / (r + 1)
  x <- qnorm(1 - alpha / 2) - abs(theta) /
    (sigma * sqrt(1 / (r * n2) + 1 / n2))
  pow <- 1 - pnorm(x)
  return(pow)
```

```
}

ggplot() + xlim(0.1, 5) +
  geom_function(
    fun = pow_ztest,
    args = list(
      n = 25,
      alpha = 0.05,
      theta = 1,
      sigma = 1
    ),
    aes(linetype = "N = 25")
  ) +
  geom_function(
    fun = pow_ztest,
    args = list(
      n = 50,
      alpha = 0.05,
      theta = 1,
      sigma = 1
    ),
    aes(linetype = "N = 50")
  ) +
  ylab("Power") + xlab(TeX("$\\r$")) +
  guides(linetype = guide_legend(title = "Allocation ratio"))
```

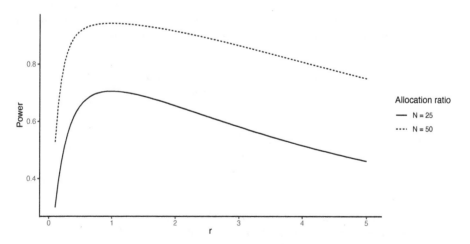

FIGURE 4.4: Power of z-test as a Function of Allocation Ratio and Sample Size

As the allocation ratio increases the power decreases assuming other parameters remain fixed (see Figure 4.4).

4.6 Power and Sample Size for Comparing Proportions

This section discusses power and sample size for studies where the primary study outcomes are dichotomous.

Let p_1 denote the response rate in group 1, p_2 the response rate in group 2, and let the difference be $\theta = p_1 - p_2$. A binary outcome for subject i in arm k is

$$Y_{ik} = \begin{cases} 1 & \text{with probability } p_k \\ 0 & \text{with probability } 1 - p_k, \end{cases}$$

for $i = 1, ..., n_k$ and $k = 1, 2$. The sum of independent and identically distributed Bernoulli random variables has a binomial distribution,

$$\sum_{i=1}^{n_k} Y_{ik} \sim Bin(n_k, p_k), \; k = 1, 2.$$

The sample proportion for group k is

$$\hat{p}_k = \frac{1}{n_k} \sum_{i=1}^{n_k} Y_{ik}, \; k = 1, 2,$$

and $E(\hat{p}_k) = p_k$ and $Var(\hat{p}_k) = p_k(1 - p_k)/n_k$.

Consider a study where the aim is to determine if there is a difference between the two groups. That is we want to test $H_0 : \theta = 0$ versus $H_1 : \theta \neq 0$.

If H_0 is true then the test statistic

$$Z = \frac{\hat{p}_1 - \hat{p}_2}{\sqrt{p_1(1 - p_1)/n_1 + p_2(1 - p_2)/n_2}} \sim N(0, 1).$$

The test rejects at level α if and only if

$$|Z| \geq z_{1-\alpha/2}.$$

Using the same argument as the case with continuous endpoints in Section 4.5.2 and ignoring terms smaller than $\alpha/2$, we can solve for β

$$\beta \approx \Phi\left(z_{1-\alpha/2} - \frac{|\theta_1|}{\sqrt{p_1(1-p_1)/n_1 + p_2(1-p_2)/n_2}}\right). \tag{4.7}$$

A formula for sample size can be derived using (4.7). If $n_1 = r \cdot n_2$ then

$$n_2 = \frac{\left(z_{1-\alpha/2} + z_{1-\beta}\right)^2}{\theta^2}\left(p_1(1-p_1)/r + p_2(1-p_2)\right),$$

where $\alpha, \beta \in (0,1)$.

4.6.1 Computation Lab: Power and Sample Size for Comparing Proportions

`stats::power.prop.test()` can be used to calculate sample size or power.

Example 4.3. The standard treatment for a disease has a response rate of 20%, and an experimental treatment is anticipated to have a response rate of 28%. The investigators want both groups to have an equal number of subjects. How many patients should be enrolled if the study will conduct a two-sided test at the 5% level with 80% power?

```
power.prop.test(p1 = 0.2, p2 = 0.28, power = 0.8)
```

```
##
##      Two-sample comparison of proportions power calculation
##
##               n = 446.2
##              p1 = 0.2
##              p2 = 0.28
##       sig.level = 0.05
##           power = 0.8
##     alternative = two.sided
##
## NOTE: n is number in *each* group
```

This means that 446 patients should be enrolled in each group for a study to have a power of 0.8 to detect $p_1 = 0.2$ and $p_2 = 0.28$ at the $\alpha = 0.05$ level.

4.7 Calculating Power by Simulation

Statistical power is the probability that the test correctly rejects the null hypothesis. Consider a thought experiment: imagine being able to replicate a study with different random samples a large number of times, and each time the null hypothesis is tested. If the null hypothesis is false, it's reasonable to expect a large proportion of these tests reject the null hypothesis. This thought experiment can be operationalized via simulation.

4.7.1 Algorithm for Simulating Power

1. Use an underlying model to generate random data with (a) specified sample sizes, (b) parameter values that are being tested via the hypothesis test, and (c) other (nuisance) parameters such as variances.

2. Run an estimation program (e.g., a two-sample t-test) on these randomly generated data.

3. Calculate the test statistic and p-value.

4. Repeat steps 1–3 many times, say, N, and save the p-values. The estimated power for a level alpha test is the proportion of observations (out of N) for which the p-value is less than alpha.

4.7.2 Computation Lab: Simulating Power of a Two-Sample t-test

If the test statistic and distribution of the test statistic are known, then the power of the test can be calculated via simulation.

Example 4.4. Consider a two-sample t-test with 30 subjects per group and the standard deviation known to be 1. What is the power of the test $H_0 : \mu_1 - \mu_2 = 0$ versus $H_1 : \mu_1 - \mu_2 = 0.5$, at the 5% significance level?

Power is the proportion of times that the test correctly rejects the null hypothesis in repeated testing of the same hypothesis based on different randomly drawn samples.

A single study is simulated below. Let's assume that $n_1 = n_2 = 30$, $\mu_1 = 3.5$, $\mu_2 = 3$, $\sigma = 1$, and $\alpha = 0.05$.

```
set.seed(2301)

samp1 <- rnorm(30, mean = 3.5, sd = 1)
samp2 <- rnorm(30, mean = 3, sd = 1)

twosampt <- t.test(samp1, samp2, var.equal = T)
twosampt$p.value
```

```
## [1] 0.03605
```

The null hypothesis would be rejected at the 5% level.

Suppose that 10 studies are simulated. What proportion of these 10 studies will reject the null hypothesis at the 5% level? To investigate how many times the two-sample t-test will reject at the 5% level, the `replicate()` function will be used to generate 10 studies and calculate the p-value in each study. It will still be assumed that $n_1 = n_2 = 30$, $\mu_1 = 3.5$, $\mu_2 = 3$, $\sigma = 1$, and $\alpha = 0.05$.

```
set.seed(2301)

tpval <- function() {
  n <- 30
  mu1 <- 3.5
  mu2 <- 3.0
  sigma <- 1
  samp1 <- rnorm(n, mean = mu1, sd = sigma)
  samp2 <- rnorm(n, mean = mu2, sd = sigma)
  twosamp <- t.test(samp1, samp2, var.equal = T)
  return(twosamp$p.value)
}

reps <- 10

pvals <- replicate(reps, tpval())
#power is the number of times the test rejects at the 5% level
sum(pvals <= 0.05) / reps
```

```
## [1] 0.3
```

But, since we only simulated 10 studies the estimate of power will have a large standard error. So let's try simulating 10,000 studies so that we can obtain a more precise estimate of power.

```
set.seed(2301)
```

```
reps <- 10000
pvals <- replicate(reps, tpval())
sum(pvals <= 0.05) / reps
```

```
## [1] 0.4881
```

This is much closer to the theoretical power obtained from stats::power.t.test().

```
powt <- power.t.test(
  n = 30,
  delta = 0.5,
  sd = 1,
  sig.level = 0.05
)
powt$power
```

```
## [1] 0.4778
```

Example 4.5. Suppose that the standard treatment for a disease has a response rate of 20%, and an experimental treatment is anticipated to have a response rate of 28%. The researchers are considering enrolling 1,500 patients in the standard group and 500 patients in the experimental arm. What is the power of this study?

The number of subjects in the experimental arm that have a positive response to treatment will be an observation from a $Bin(1500, 0.20)$ and the number of subjects that have a positive response to the standard treatment will be an observation from a $Bin(500, 0.28)$. We can obtain simulated responses from these distributions using the rbinom() function.

```
set.seed(2301)
rbinom(1, 500, 0.28)
```

```
## [1] 132
```

```
rbinom(1, 1500, 0.20)
```

```
## [1] 324
```

In this simulated study, 132 of the 500 patients in the experimental arm had a positive response to the experimental treatment and 324 of the 1,500 patients in the control arm had a positive response to the standard treatment. The p-value for this simulated study can be obtained using prop.test().

```
set.seed(2301)

samp1 <- rbinom(1, 500, 0.28)
samp2 <- rbinom(1, 1500, 0.20)

twosamp <- prop.test(
  x = c(samp1, samp2),
  n = c(500, 1500),
  correct = F
)

twosamp$p.value
```

```
## [1] 0.02672
```

In this study, the p-value is 0.03, which is less than 0.05 so there would be evidence that the new treatment is significantly better than the standard treatment. A power simulation repeats this process a large number of times. The replicate() command can be used for the repetition.

```
set.seed(2301)

proppval <- function() {
  n1 <- 500
  p1 <- 0.28
  n2 <- 1500
  p2 <- 0.20
  samp1 <- rbinom(n = 1, size = n1, prob = p1)
  samp2 <- rbinom(n = 1, size = n2, prob = p2)
  twosamp <- prop.test(x = c(samp1, samp2),
                       n = c(n1, n2),
```

```
                            correct = F)
   return(twosamp$p.value)
}

reps <- 10000
pvals <- replicate(reps, proppval())
sum(pvals <= 0.05) /  reps
```

[1] 0.9534

The power of the study in this case is 0.95.

4.8 Exercises

Exercise 4.1. Consider the power function of one-sample z-test shown in Equation (4.3). What is the limit of the power function as $n \to \infty$? How about when $\mu_1 \to \mu_0$? What do the results tell you about the power of the one-sample z-test?

Exercise 4.2. Show that

$$P_{H_1}\left(\frac{\bar{X} - \mu_0}{\frac{S}{\sqrt{n}}} \geq t_{n-1,1-\alpha/2}\right) + P_{H_1}\left(\frac{\bar{X} - \mu_0}{\frac{S}{\sqrt{n}}} < -t_{n-1,1-\alpha/2}\right)$$
$$= P(t_{n-1,\gamma} \geq t_{n-1,1-\alpha/2}) + P(t_{n-1,\gamma} < -t_{n-1,1-\alpha/2}).$$

The right-hand side is the final expression for the power of the t-test in Section 4.4.

Exercise 4.3. Derive equation (4.1) for the total sample of the two-sample t-test.

Exercise 4.4. Consider the derivation of Equation (4.5) for the sample size of a two-sample z-test with known variance and equal allocation.

a. Identify terms in Equation (4.2) that are smaller than $\alpha/2$ and derive line (4.3) based on the information.

b. Show how Equation (4.4) is derived from Equation (4.3).

Exercise 4.5. Consider Figure 4.2 and the R code that generated the plots.

a. Modify the code so that sample size is on the x-axis, and three different lines show the relationships between sample size and power for a difference of $\theta = 1, 2, 3$. Fix the standard deviation to 1.5 for all lines.

b. Interpret the relationships between power, effect size θ, standard deviation σ, and sample size n, on Figure 4.2 and the figure from part a. For example, what happens to power as θ decreases with n and σ fixed? How does θ change as power increases with other parameters fixed?

Exercise 4.6. Consider the plot of power as a function of effect size for the two-sample t-test shown in Figure 4.3.

a. Create a plot to show power as a function of both effect size and sample size while keeping other parameters fixed.

b. The curve is shaped like an inverted bell curve, or a "bathtub" shape. Interpret the shape of the curve.

Exercise 4.7. Consider the plot of power as a function of the allocation ratio, r, for the two-sample z-test shown in Figure 4.4 and the R code for generating plot.

a. Recall that n is the total sample size for the function `pow_ztest()`. What does n2 represent, and why is it n/(r+1)?

b. Does sample size imbalance between the two groups lead to an increase or decrease in power? Explain.

c. Modify the function `pow_ztest()` to return the value of n2 and plot n2 versus r. Describe the relationship between the two variables.

Exercise 4.8. The R function `size2z.test()` shown below implements the sample size formula for calculating the sample size for a test of $H_0 : \theta = 0$ versus $H_1 : \theta \neq 0$, where $\theta = \mu_1 - \mu_2$.

```
size2z.test <- function(theta, alpha, beta, sigma)
{
  zalpha <- qnorm(1-alpha/2)
  zbeta <- qnorm(1-beta)
  (2*sigma*(zalpha+zbeta)/theta)^2
}
```

a. What are the main assumptions behind this sample size formula?

b. Assuming all other parameters remain fixed, indicate whether the sample size increases or decreases in each of the following cases.

    i. As σ decreases.

   ii. As α decreases.

  iii. As θ decreases.

Exercise 4.9. A statistician is designing a phase III clinical trial comparing a continuous outcome in two groups receiving experimental versus standard therapy with a total sample size of 168 patients. The team requires the study have 80% power at the 5% significance level to detect a difference of 1. Assume that the standard deviation of the outcome is 2. The design team would like to investigate whether it's possible to have four times as many patients in the experimental group versus the control group without having to increase the total sample size.

What is the power if there are four times as many patients in the experimental group? What should the statistician recommend to the team in order for the study to have at least 80% power?

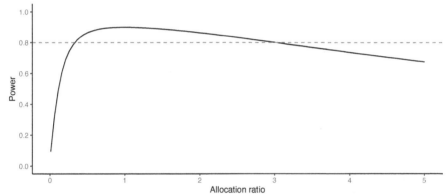

Exercise 4.10. Let $X_1, X_2, ..., X_n$ be iid $N(\mu, \sigma^2)$.

a. Show that the power function of the test $H_0 : \mu = 0$ versus $H_1 : \mu > 0$ at $\mu = 1$ is

$$1 - \Phi\left(z_{\frac{\alpha}{2}} - \frac{\sqrt{n}}{\sigma}\right),$$

where $z_{\frac{\alpha}{2}}$ is the $100\left(1 - \frac{\alpha}{2}\right)^{th}$ percentile of the $N(0, 1)$.

b. Use R to calculate the power when $n = 10$, $\alpha = 0.01$, and $\sigma = 1$.

5

Comparing More Than Two Treatments

5.1 Introduction: ANOVA—Comparing More Than Two Groups

Chapter 3 discussed design and analysis of comparing two treatments by randomizing experimental units to two different treatment groups. This chapter discusses design and analysis for comparing more than two treatments.

Example 5.1 (Three-arm randomized clinical trial). Amundson et al. [2017] conducted a clinical trial of three different treatments to control pain during knee surgery. The primary outcome of the study was maximal pain (0 to 10, numerical pain rating scale where higher values indicate more severe pain) one day after surgery. The study aimed to enrol 150 participants randomized in equal numbers to the three treatments (in clinical trials, treatments are often called arms—short for treatment arms).

TABLE 5.1: Example 5.1 Treatment Means and Standard Deviations

trt	N	Mean	SD
A	50	6.20	1.34
B	50	7.32	1.15
C	50	3.64	1.51

Is there evidence to indicate a difference in mean pain scores for the three different pain treatments? An idea due to R.A. Fisher is to compare the variation in mean pain scores *between* the treatments to the variation of pain scores *within* a treatment. These two measures of variation are often summarized in an **analysis of variance (ANOVA) table**.

5.2   Random Assignment of Treatments

The number of possible randomizations in Example 5.1 is

$$\binom{150}{50\,50\,50} = \frac{150!}{50!50!50!} \approx 2.03 \times 10^{69}.$$

One of the random allocations could be selected by taking a random permutation of the numbers 1 through 150 and then assigning the first 50 to the first treatment, the second 50 to the second treatment, etc. One of the drawbacks of this method is that the sample size in each group will be unequal until all the participants are randomized. In studies such as clinical trials, it's common to plan an *interim analysis* before all the participants are enrolled to evaluate safety and efficacy. If the groups are unbalanced, then this could lead to a loss of power (see Chapter 4).

Permuted block randomization avoids this problem by randomizing experimental units within a smaller block of experimental units. The order of treatment assignments within a block are random, but the desired proportion of treatments is achieved in each block. In Example 5.1 participants could be randomized to treatments in a block size of 6: 2 assigned to treatment 1; 2 assigned to treatment 2; and 2 assigned to treatment 3. If the treatments are labeled A, B, and C then possible blocks could be: ABBACC, CABBCA, or CBAACB. As the blocks are filled, the study is guaranteed to have the desired allocation to each group.

5.2.1   Computation Lab: Random Assignment

A randomization scheme can be created for Example 5.1 by labeling participants with the sequence 1:150, selecting a random permutation, then assigning the first 50 to treatment A, etc.

```
set.seed(1)
subj <- sample(1:150)
trt <- c(rep("A", 50), rep("B", 50), rep("C", 50))
randscheme <- data.frame(subj, trt) %>% arrange(subj, trt)
head(randscheme)
```

```
##    subj trt
## 1     1   B
```

```
## 2      2    A
## 3      3    C
## 4      4    C
## 5      5    C
## 6      6    A
```

Participant 1 is randomized to treatment B, etc. The treatment assignments to the first 6 participants indicate an unequal number of treatments since there are 2 A's, 3 C's, and 1 B.

A permuted block randomization will correct this issue. Consider blocks of size 6. To obtain a random order of treatments in block sizes of 6, we can replicate `trts` twice and then `sample()` the resulting vector to obtain a random permutation. The function `create_blocks()` creates blocks of size `m` × `length(trts)`.

```
trts <- c("A", "B", "C")

create_block <- function(m) {
  sample(rep(trts, m))
}
```

To create a randomization scheme with a block size of 6 for the study, we can run `create_block()` $150/6 = 25$ times using `replicate()`.

```
randscheme <- replicate(25, create_block(m = 2))
```

5.3 ANOVA

5.3.1 Source of Variation and Sums of Squares

Let $y_{ij}, i = 1, \ldots, k; j = 1, \ldots n_i$ be the jth observation with treatment i. The ith treatment mean will be denoted as $\bar{y}_{i\cdot} = y_{i\cdot}/n_i$, where $y_{i\cdot} = \sum_{j=1}^{n_i} y_{ij}$. The grand (overall) mean is $\bar{y}_{\cdot\cdot} = y_{\cdot\cdot}/N$, where $y_{\cdot\cdot} = \sum_{i=1}^{k} \sum_{j=1}^{n_i} y_{ij}$. $N = \sum_{i=1}^{k} n_i$ is the total number of observations. The "dot" subscript notation means sum over the subscript that it replaces.

Consider a scenario such as Example 5.1. The between treatments variation and within treatment variation are two components of the total variation in the response. We can break up each observation's deviation from the grand mean $y_{ij} - \bar{y}_{..}$ into two components: treatment deviations $(\bar{y}_{i.} - \bar{y}_{..})$ and residuals within treatment deviations $(y_{ij} - \bar{y}_{i.})$.

$$y_{ij} - \bar{y}_{..} = (\bar{y}_{i.} - \bar{y}_{..}) + (y_{ij} - \bar{y}_{i.}),$$

where y_{ij} is the jth observation, $j = 1, \ldots, n_i$ taken under treatment $i = 1, ..., k$.

The statistical model that we will study in this section is

$$y_{ij} = \mu + \tau_i + \epsilon_{ij}, \tag{5.1}$$

where τ_i is the ith treatment effect, and $\epsilon_{ij} \sim N(0, \sigma^2)$ is the random error of the jth observation in the ith treatment. For example, the average of the observations under treatment one $(i = 1)$, $E(y_{1j})$, minus the grand average, μ, $\tau_1 = E(y_{1j}) - \mu$ is the first treatment effect.

We are interested in testing whether the k treatment effects are equal.

$$H_0 : \tau_1 = \cdots = \tau_k \quad \text{vs.} \quad H_1 : \tau_l \neq \tau_m, \, l \neq m.$$

5.3.2   ANOVA Identity

The total sum of squares $SS_T = \sum_{i=1}^{k} \sum_{j=1}^{n_i} (y_{ij} - \bar{y}_{..})^2$ can be written as

$$\sum_{i=1}^{k} \sum_{j=1}^{n_i} [(\bar{y}_{i.} - \bar{y}_{..}) + (y_{ij} - \bar{y}_{i.})]^2$$

by adding and subtracting $\bar{y}_{i.}$ to $(y_{ij} - \bar{y}_{..})$.

Expanding and summing the expression leads to

$$SS_T = \underbrace{\sum_{i=1}^{k} \sum_{j=1}^{n_i} (y_{ij} - \bar{y}_{..})^2}_{} = \underbrace{\sum_{i=1}^{k} n_i (\bar{y}_{i.} - \bar{y}_{..})^2}_{\text{Sum of Squares Due to Treatment}} + \underbrace{\sum_{i=1}^{k} \sum_{j=1}^{n_i} (y_{ij} - \bar{y}_{i.})^2}_{\text{Sum of Squares Due to Error}} \tag{5.2}$$

$$= SS_{Treat} + SS_E. \tag{5.3}$$

This is sometimes called the analysis of variance identity. It shows how the total sum of squares can be split into two sums of squares: one part that is due to differences between treatments, and one part due to differences within treatments. The squared deviations SS_{Treat} are called the sum of squares due to treatments (i.e., between treatments), and SS_E is called the sum of squares due to error (i.e., within treatments).

5.3.3   Computation Lab: ANOVA Identity

R functions to compute SS_T, SS_{Treat}, and SS_E are below. SST() computes the sum of squared deviations: SSTreat() first **splits** the data frame into treatment **groups** then computes the sum of squared deviations of the treatment means sst multiplied by n the number of observations within a treatment group, and finally SSe() computes the sum of squared deviations of each observation within a treatment from its treatment mean. lapply() is used in SSTreat() and SSe() to compute the function over a vector or list and returns a list, unlist() converts the list to a vector so that we can apply sum().

```
SST <- function(y) {
  sum((y - mean(y)) ^ 2)
}

SSTreat <- function(y, groups, n) {
  sst <- unlist(lapply(split(y, groups),
                       function(x) {
                         (mean(x) - mean(y)) ^ 2
                       }))
  return(n * sum(sst))
}

SSe <- function(y, groups) {
  sum(unlist(lapply(split(y, groups),
                    function(x) {
                      (x - mean(x)) ^ 2
                    })))
}
```

The **painstudy** data frame contains data for Example 5.1, and the ANOVA identity (5.3) can be verified using the functions above.

5.3.4   ANOVA: Degrees of Freedom

Suppose you are asked to choose a pair of real numbers (x, y) at random. This means that you have complete freedom to choose the two numbers, or two *degrees of freedom*. Now, suppose that you are asked to choose a pair of real numbers (x, y), such that $x + y = 10$. Once you choose, say x, then y is fixed, so there is only one *degree of freedom* in this case.

What are the degrees of freedom of SS_T? There are $N = \sum_i^k n_i$ *residual* terms $(y_{ij} - \bar{y}..)$, such that $\sum_{i=1}^{k} \sum_{j=1}^{n_i} (y_{ij} - \bar{y}..) = 0$. This means that any $N - 1$ are completely determined by the other. Therefore, SS_T has $N - 1$ degrees of freedom. Similarly, SS_{Treat} has $k - 1$ degrees of freedom. Within each treatment there are n_i observations with $n_i - 1$ degrees of freedom, and there are k treatments. So, there are $\sum_{i=1}^{k}(n_i - 1) = N - k$ degrees of freedom for SS_E.

5.3.5   ANOVA: Mean Squares

$$SS_E = \sum_{i=1}^{k} \left[\sum_{j=1}^{n_i} (y_{ij} - \bar{y}_{i.})^2 \right]$$

If the term inside the brackets is divided by $n - 1$, then it is the sample variance for the *ith* treatment

$$S_i^2 = \frac{\sum_{j=1}^{n_i} (y_{ij} - \bar{y}_{i.})^2}{(n_i - 1)}, \qquad 1 = 1, ..., k.$$

Combining these k variances to give a single estimate of the common population variance,

$$\frac{(n_1 - 1)S_1^2 + \cdots + (n_k - 1)S_k^2}{(n_1 - 1) + \cdots + (n_k - 1)} = \frac{SS_E}{N - k}.$$

It can be shown that $E(SS_E) = (N - k)\sigma^2$. So, $SS_E/(N - k)$ can be used to estimate σ^2.

If there were no differences between the k treatment means $\bar{y}_i$. we could use the variation of the treatment averages from the grand average to estimate σ^2.

$$\frac{SS_{Treat}}{k - 1}.$$

It can be shown that $E(SS_{Treat}) = k \sum_{i=1}^{k} \tau_i^2 + (k - 1)\sigma^2$. So, if $\tau_i = 0$, then $SS_{Treat}/(k - 1)$ is an estimate of σ^2 when the treatment deviations are all equal.

The analysis of variance identity can be used to give two estimates of σ^2: one based on the variability within treatments and the other based on the variability between treatments. These two estimators of σ^2 are called mean square treatment and error.

Definition 5.1 (Mean Square for Treatment). The **mean square for treatment** is defined as
$$MS_{Treat} = \frac{SS_{Treat}}{k-1}.$$

Definition 5.2 (Mean Square for Error). The **mean square for error** is defined as
$$MS_E = \frac{SS_E}{N-k}.$$

5.3.6   ANOVA: F Statistic

It can be shown that SS_{Treat} and SS_E are independent. If $H_0 : \tau_1 = \cdots = \tau_k = 0$ is true, then $SS_{Treat}/\sigma^2 \sim \chi^2_{k-1}$ and $SS_E/\sigma^2 \sim \chi^2_{N-k}$, so
$$F = \frac{MS_{Treat}}{MS_E} \sim F_{k-1,N-k}.$$

5.3.7   ANOVA Table

A table that records the source of variation, degrees of freedom, sum of squares, mean squares, and F statistic values is called an ANOVA table.

TABLE 5.2: ANOVA Table

Source of variation	Degrees of freedom	Sum of squares	Mean square	F
Between treatments	$k-1$	SS_{Treat}	MS_{Treat}	F
Within treatments	$N-k$	SS_E	MS_E	

Where, $F = \frac{MS_{Treat}}{MS_E}$.

5.3.8   Computation Lab: ANOVA Table

The `aov()` function fits an ANOVA model and the `summary()` function computes the ANOVA table. The ANOVA table for Example 5.1 is

```
anova_mod <- aov(pain ~ trt, data = painstudy)
summary(anova_mod)
```

```
##              Df Sum Sq Mean Sq F value Pr(>F)
## trt           2    356   177.9    98.9 <2e-16 ***
## Residuals   147    264     1.8
## ---
## Signif. codes:
## 0 '***' 0.001 '**' 0.01 '*' 0.05 '.' 0.1 ' ' 1
```

5.4 Estimating Treatment Effects Using Least Squares

The statistical model for ANOVA (5.1) can also be described as a linear regression model (2.4).

5.4.1 Dummy Coding

In Example 5.1, y_{ij} is the jth pain score, $j = 1, \ldots, 50$ for the ith treatment group, $i = 1, 2, 3$, which correspond to treatments A, B, and C, respectively.

TABLE 5.3: First Six Observations of Example 5.1

A	B	C
6	8	3
7	5	1

Let $\mathbf{y} = \underbrace{(y_{11}, y_{21}, \ldots, y_{50\,3})'}_{150 \times 1}$ represent the column vector $(6, 8, 3, \ldots, 5, 9, 6)'$,

$$X = \begin{pmatrix} 1 & x_{11} & x_{12} \\ 1 & x_{21} & x_{22} \\ \vdots & \vdots & \vdots \\ 1 & x_{150\,1} & x_{150\,2} \end{pmatrix}, \quad \beta = \begin{pmatrix} \mu \\ \tau_1 \\ \tau_2 \end{pmatrix}, \quad \epsilon = \begin{pmatrix} \epsilon_{11} \\ \epsilon_{21} \\ \vdots \\ \epsilon_{3\,50} \end{pmatrix}$$

where,

$$x_{1j} = \begin{cases} 1 & \text{if jth unit receives treatment B} \\ 0 & \text{otherwise} \end{cases},$$

$$x_{2j} = \begin{cases} 1 & \text{if jth unit receives treatment C} \\ 0 & \text{otherwise} \end{cases}.$$

This regression model implicitly assumes that the regression parameter for treatment A is zero (e.g., $\tau_3 = 0$). This is often called the **baseline constraint**.

The predictors x_{1j} and x_{2j} are coded as **dummy variables**; **dummy coding** compares each level to a reference level, and the intercept is the mean of the reference group. In this example, we have used two categorical covariates, coded as dummy variables, to describe three treatments (see Section 2.8.3); otherwise the regression model would be overparameterized and $(X'X)^{-1}$ would not exist.

We can also write the regression model using (2.3)

$$y_{ij} = \mu + \tau_1 x_{1j} + \tau_2 x_{2j} + \epsilon_{ij}. \tag{5.4}$$

If $i = 3$, then $E(y_{3j}) = \mu$, the intercept of this model, is the mean pain score of the treatment A group. Letting $\mu = \mu_A$ and using a similar notation for the other treatments:

$$E(y_{1j}) = \mu_B = \mu + \tau_1 \Rightarrow \tau_1 = \mu_B - \mu_A,$$
$$E(y_{2j}) = \mu_C = \mu + \tau_2 \Rightarrow \tau_2 = \mu_C - \mu_A.$$

Using (2.5) we can obtain the least squares estimates:

$$\hat{\mu} = \bar{y}_{1.}, \quad \hat{\tau}_1 = \bar{y}_{2.} - \bar{y}_{1.}, \quad \hat{\tau}_2 = \bar{y}_{3.} - \bar{y}_{1.}.$$

5.4.2 Deviation Coding

This coding system compares the mean of the dependent variable for a given level to the overall mean of the dependent variable. Suppose the independent variables in model (5.4) are defined as

$$x_{1j} = \begin{cases} 1 & \text{if treatment A} \\ -1 & \text{if treatment C} \\ 0 & \text{otherwise} \end{cases}, \quad x_{2j} = \begin{cases} 1 & \text{if treatment B} \\ -1 & \text{if treatment C} \\ 0 & \text{otherwise} \end{cases}$$

This coding is equivalent to assuming a regression model where $\tau_1 + \tau_2 + \tau_3 = 0$, where τ_3 is the regression parameter for treatment C. This is often called the **zero-sum constraint**. It is another type of constraint on the regression model to ensure that $X'X$ is non-singular.

1 is used to compare a level to all other levels and -1 is assigned to treatment C because it's the level that will never be compared to the other levels.

Let $i = 1$, so $\mu_A = E(y_{1j})$. Then $\mu_A = \mu + \tau_1$ is the mean pain score of the treatment A group. Continuing with this notation, we have $\mu_B = \mu + \tau_2$, and $\mu_C = \mu - \tau_1 - \tau_2$. Solving for μ, τ_1, and τ_2, we have

$$\mu = \frac{\mu_A + \mu_B + \mu_C}{3}, \quad \tau_1 = \mu_A - \mu, \quad \tau_2 = \mu_B - \mu.$$

5.5   Computation Lab: Estimating Treatment Effects Using Least Squares

lm() can be used to fit a linear regression model that calculates estimates of the ANOVA treatment effects. In this section, we will use the data in Example 5.1. The dependent variable is **pain** and the independent variable is **trt**. But, let's start with checking the coding of **trt** using stats::contrasts().

painstudy$trt is a character vector, but contrasts() takes a factor vector as an argument. So, we can use as.factor() to convert to an R factor object.

```
painstudy$trt <- as.factor(painstudy$trt)
contrasts(painstudy$trt)
```

```
##   B C
## A 0 0
## B 1 0
## C 0 1
```

The output shows a matrix whose columns define dummy variables x_{1j} and x_{2j} with treatment A as the reference category (since it has a row of 0's). This coding (i.e., contrast) is the default in R for unordered factors, and can be specified using contr.treatment(). To change the reference category, use the base option in contr.treatment().

```
contrasts(painstudy$trt) <- contr.treatment(n = 3, base = 2)
contrasts(painstudy$trt)
```

```
##   1 3
## A 1 0
## B 0 0
```

```
## C 0 1
```

contr.treatment(n = 3, base = 2) indicates that there are n = 3 treatment levels, and the reference category is the second category base = 2 (the
default is base = 1). Contrast can also be specified in lm().

```
aovregmod <- lm(pain ~ trt,
                data = painstudy,
                contrasts =
                  list(trt = contr.treatment(n = 3, base = 2)))
broom::tidy(aovregmod)
```

```
## # A tibble: 3 x 5
##    term         estimate std.error statistic  p.value
##    <chr>           <dbl>     <dbl>     <dbl>    <dbl>
## 1 (Intercept)      7.32     0.190      38.6  8.17e-79
## 2 trt1            -1.12     0.268      -4.18 5.07e- 5
## 3 trt3            -3.68     0.268     -13.7  4.22e-28
```

Table 5.1 gives the treatment means. The estimate of treatment B mean is
the (Intercept) term (7.32); the estimate for trt1 is the difference between
the means for treatments A (6.2) and B (7.32); and the estimate for trt3
corresponds to the difference between the means for treatments C (3.64) and
B (7.32).

```
confint(aovregmod)
```

```
##                2.5 %  97.5 %
## (Intercept)   6.945   7.6948
## trt1         -1.650  -0.5899
## trt3         -4.210  -3.1499
```

Confidence intervals for the treatment estimates can be obtained using the
confint() function. Neither confidence interval for the treatment effects
contain zero so would be considered statistically significant at the 5% level (this
can also be ascertained by examining the p-values in summary(aovregmod)).

Alternatively, we could use deviation coding by specifying the contrast as
contr.sum().

```
contrasts(painstudy$trt) <- contr.sum(3)
contrasts(painstudy$trt)
```

```
##    [,1] [,2]
## A    1    0
## B    0    1
## C   -1   -1
```

In this contrast, treatment C is never compared to the other treatments. If we wish to change this to, say, treatment A, then use the `levels` argument of `factor()` to reorder the levels of a factor.

```
painstudy$trt <- factor(painstudy$trt, levels = c("C", "B", "A"))
contrasts(painstudy$trt) <- contr.sum(3)
contrasts(painstudy$trt)
```

```
##    [,1] [,2]
## C    1    0
## B    0    1
## A   -1   -1
```

Now, we can fit the regression model with deviation coding where treatment A is the comparison treatment.

```
aovregmod <- lm(pain ~ trt, data = painstudy)
broom::tidy(aovregmod)
```

```
## # A tibble: 3 x 5
##   term        estimate std.error statistic  p.value
##   <chr>          <dbl>     <dbl>     <dbl>    <dbl>
## 1 (Intercept)     5.72     0.110      52.2 8.13e-97
## 2 trt1           -2.08     0.155     -13.4 2.42e-27
## 3 trt2            1.60     0.155      10.3 3.77e-19
```

The (Intercept) is the grand mean or mean of means ($6.2 + 7.32 + 3.64$)/3 = 5.72; `trt1` is the mean of treatment C 3.64 minus the the grand mean, and `trt2` is the mean of treatment B 7.32 minus the grand mean.

5.5.1 ANOVA Assumptions

The calculations that make up an ANOVA table require no assumptions. You could complete the table using the ANOVA identity and definitions of mean square and F statistic for a data set with k treatments. However, using these numbers to make inferences about differences in treatment means requires certain statistical assumptions.

1. Additive model. The observations y_{ij} can be modelled as the sum of treatment effects and random error, $y_{ij} = \mu + \tau_i + \epsilon_{ij}$. The parameters τ_i are interpreted is the ith treatment effect.

2. Under the assumption that the errors ϵ_{ij} are independent and identically distributed with a common variance $Var(\epsilon_{ij}) = \sigma^2$, for all i, j, $E(MS_{Treat}) = (k/k-1)\sum_{i=1}^{k} \tau_i^2 + \sigma^2$, and $E(MS_E) = \sigma^2$. If $\tau_i = 0, i = 1, \ldots, k$ then $E(MS_{treat}) = E(MS_E) = \sigma^2$.

3. If $\epsilon_{ij} \sim N(0, \sigma^2)$ then MS_{Treat} and MS_E are independent. Under the null hypothesis, $H_0 : \tau_i = 0, i = 1, \ldots, a$, the ratio $F = MS_{Treat}/MS_E \sim F_{a-1, N-a}$.

5.5.2 Computation Lab: Checking ANOVA Assumptions

In this section, we check the ANOVA assumptions (Section 5.5.1) for Example 5.1.

1. The ANOVA decomposition implies acceptance of the additive model $y_{ij} = \mu + \tau_j + \epsilon_{ij}$, where μ is the mean of treatment A and τ_j is the deviation produced by treatment j, and ϵ_{ij} is the associated random error. There are no formal methods that we have introduced to check this assumption.

2. The common variance assumption can be investigated by plotting the residuals versus the fitted values of the ANOVA model. A plot of the residuals versus fitted values can be used to investigate the assumption that the residuals are randomly distributed and have constant variance. Ideally, the points should randomly fall on both sides of 0, with no recognizable patterns in the points. In R, this can be done using the following commands.

```
data.frame(resid = aovregmod$residuals,
           pred = aovregmod$fitted.values) %>%
  ggplot(aes(x = pred, y = resid)) +
```

```
geom_point() +
geom_hline(yintercept = 0)
```

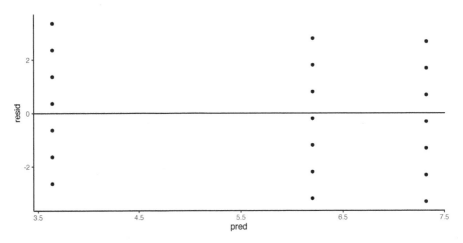

The assumption of constant variance is satisfied since there is no recognizable pattern.

3. The normality of the residuals can be investigated using a normal quantile-quantile plot (see 2.5).

5.6   Multiple Comparisons

Suppose that experimental units were randomly assigned to three treatment groups. The hypothesis of interest is

$$H_0 : \mu_1 = \mu_2 = \mu_3 \text{ vs. } H_1 : \mu_i \neq \mu_j.$$

If H_0 is rejected at level α, then which pairs of means are significantly different from each other at level α? There are $\binom{3}{2} = 3$ possibilities.

1. $\mu_1 \neq \mu_2$
2. $\mu_1 \neq \mu_3$
3. $\mu_2 \neq \mu_3$

Suppose that $k = 3$ separate (independent) hypothesis level α tests are conducted:

$$H_{0_k} : \mu_i = \mu_j \text{ vs. } H_{1_k} : \mu_i \neq \mu_j.$$

When H_0 is true, $P_{H_0}(\text{reject } H_0) = \alpha \Rightarrow P_{H_0}(\text{not reject } H_0) = 1 - \alpha$. So, if H_0 is true, then

$$
\begin{aligned}
P_{H_0}\left(\text{reject at least one } H_{0_k}\right) &= 1 - P_{H_0}\left(\text{not reject any } H_{0_k}\right) \\
&= 1 - P_{H_0}\left(\text{not reject } H_{0_1} \text{and not reject } H_{0_2} \text{and not reject } H_{0_3}\right) \\
&= 1 - P_{H_0}\left(\text{not reject } H_{0_1}\right) P\left(\text{not reject } H_{0_2}\right) P\left(\text{not reject } H_{0_3}\right) \\
&= 1 - (1 - \alpha)^3
\end{aligned}
$$

If $\alpha = 0.05$ then the probability that at least one H_0 will be falsely rejected is $1 - (1 - .05)^3 = 0.14$, which is almost three times the type I error rate.

In general if
$$H_0 : \mu_1 = \mu_2 = \cdots = \mu_k \text{ vs. } H_1 : \mu_i \neq \mu_j,$$

and $c = \binom{k}{2}$ independent hypotheses are conducted, then the probability

$$P_{H_0}\left(\text{reject at least one } H_{0_k}\right) = 1 - (1 - \alpha)^c$$

is called the **family-wise error rate** (FWER).

The **pairwise error rate** is $P_{H_0}\left(\text{reject } H_{0_k}\right) = \alpha$ for any c.

The *multiple comparisons problem* is that multiple hypotheses are tested at level α which increases the probability that at least one of the hypotheses will be falsely rejected (family-wise error rate).

When an ANOVA F test is significant, then investigators often wish to explore where the differences lie. Is it appropriate to test for differences looking at all pairwise comparisons when testing all possible pairs increases the type I error rate? An increased type I error rate (e.g. 0.14) means that that there is a higher probability of detecting a significant difference when no difference exists.

In the next two sections, we will cover two methods, Bonferroni and Tukey, for adjusting the p-values so that the FWER does not increase when all pairwise comparisons are conducted.

5.6.1   The Bonferroni Method

To test for the difference between the ith and jth treatments, it is common to use the two-sample t-test. The two-sample t statistic is

$$t_{ij} = \frac{\bar{y}_{j\cdot} - \bar{y}_{i\cdot}}{\hat{\sigma}\sqrt{1/n_j + 1/n_i}},$$

where $\bar{y}_{j\cdot}$ is the average of the n_i observations for treatment j and $\hat{\sigma}$ is $\sqrt{MS_E}$ from the ANOVA table.

Treatments i and j are declared significantly different at level α if

$$|t_{ij}| > t_{N-k, 1-\alpha/2}.$$

The total number of pairs of treatment means that can be tested is

$$c = \binom{k}{2} = \frac{k(k-1)}{2}.$$

The Bonferroni method for testing $H_0 : \mu_i = \mu_j$ vs. $H_0 : \mu_i \neq \mu_j$ rejects H_0 at level α if

$$|t_{ij}| > t_{N-k, \alpha/2c},$$

where c denotes the number of pairs being tested.

5.6.2   The Tukey Method

Treatments i and j are declared significantly different at level α if

$$|t_{ij}| > \frac{1}{\sqrt{2}} q_{k, N-k, \alpha},$$

where t_{ij} is the observed value of the two-sample t-statistic and $q_{k, N-k, \alpha}$ is the upper α percentile of the Studentized range distribution with parameters k and $N - k$ degrees of freedom.

A $100(1 - \alpha)\%$ simultaneous confidence interval for c pairs $\mu_i - \mu_j$ is

$$\bar{y}_{j\cdot} - \bar{y}_{i\cdot} \pm \frac{1}{\sqrt{2}} q_{k, N-k, \alpha} \hat{\sigma} \sqrt{1/n_j + 1/n_i}.$$

The Bonferroni method is more conservative than Tukey's method. In other words, the simultaneous confidence intervals based on the Tukey method are shorter.

The main difference between the Tukey and Bonferroni methods is in the choice of the critical value.

5.6.3   Computation Lab: Multiple Comparisons

Example 5.2 (Four-arm randomized clinical trial). A pain study similar to Example 5.1 was conducted, except that 120 participants were randomized to four pain treatments during knee surgery in equal numbers. The primary outcome of the study was maximal pain (0 to 10, numerical pain rating scale where higher values indicate more severe pain) one day after surgery. The goal of the study was to evaluate which treatments were effective. The data for this study is in the data frame `painstudy2`.

The library `emmeans` [Lenth, 2021] has functions that can compute estimated treatment (also called marginal) means, comparisons and contrasts with adjusted p-values and confidence intervals. `emmeans()` computes the marginal means, and `pairs()` computes all pairwise comparisons with an option to `adjust` the comparisons using `tukey` or `bonferroni`.

We first fit a regression model and then compute the means for each `trt` using `emmeans::emmeans()`.

```
library(emmeans)
painstudy2mod <- lm(pain ~ trt, data = painstudy2)
painstud.emm <- emmeans(painstudy2mod, specs = "trt")
painstud.emm
```

```
## trt emmean    SE  df lower.CL upper.CL
## A    6.60 0.237 116     6.13     7.07
## B    7.37 0.237 116     6.90     7.84
## C    3.20 0.237 116     2.73     3.67
## D    7.20 0.237 116     6.73     7.67
##
## Confidence level used: 0.95
```

To obtain all $\binom{4}{2} = 6$ pairwise comparisons with the Tukey adjustment we use the `pairs()` function with `adjust = tukey`, and using `confint()` adds confidence intervals for each comparison.

```
ci <- confint(pairs(painstud.emm, adjust = "tukey"))
```

Unadjusted 95% confidence intervals are obtained using `adjust = "none"`.

```
confint(pairs(painstud.emm, adjust = "none"))
```

```
##  contrast estimate     SE  df lower.CL upper.CL
##  A - B      -0.767 0.335 116   -1.431   -0.102
##  A - C       3.400 0.335 116    2.736    4.064
##  A - D      -0.600 0.335 116   -1.264    0.064
##  B - C       4.167 0.335 116    3.502    4.831
##  B - D       0.167 0.335 116   -0.498    0.831
##  C - D      -4.000 0.335 116   -4.664   -3.336
##
## Confidence level used: 0.95
```

Adjusting for multiple comparisons in this case has an impact on statistical signifcance at the 5% level, since the comparison `A - B` is no longer significant after the Tukey adjustment. Alternatively, the CDF and inverse CDF of the Studentized Range Distribution `stats::ptukey()` and `stats::qtukey()` could be used directly to compute the confidence intervals.

We can also obtain plots of the adjusted confidence intervals.

```
plot(painstud.emm, comparisons = TRUE, adjust = "tukey")
pwpp(painstud.emm, comparisons = TRUE, adjust = "tukey")
```

An explanation on how to interpret these plots is given by Lenth [2021].

5.7 Sample Size for ANOVA—Designing a Study to Compare More Than Two Treatments

When designing a study to compare more than two treatments, investigators often need to consider how many experimental units will be assigned to each group to detect differences if they exist. In this section the methods introduced in Chapter 4 are extended to ANOVA.

$$H_0 : \mu_1 = \mu_2 = \cdots = \mu_k \text{ vs. } H_1 : \mu_i \neq \mu_j. \tag{5.5}$$

is an ANOVA test comparing k treatment means. The test rejects at level α if

$$MS_{Treat}/MS_E \geq F_{k-1,N-K,\alpha}.$$

The power of the test is $P_{H_1}\left(MS_{Treat}/MS_E \geq F_{k-1,N-K,\alpha}\right)$. In order to compute this probability, we require the distribution of MS_{Treat}/MS_E when H_0 is false. It can be shown that MS_{Treat}/MS_E has a non-central F distribution with the numerator and denominator degrees of freedom $k - 1$ and $N - k$ respectively, and non-centrality parameter

$$\delta = \frac{\sum_{i=1}^{k} n_i \left(\mu_i - \bar{\mu}\right)^2}{\sigma^2}, \tag{5.6}$$

where n_i is the number of observations in group i, $\bar{\mu} = \sum_{i=1}^{k} \mu_i/k$, and σ^2 is the within group error variance.

Let $F_{a,b}(\delta)$ denote the non-central F distribution with numerator and denominator degrees of freedom a, b and non-centrality parameter δ.

The power of (5.5) is $P\left(F_{k-1,N-k}(\delta) > F_{k-1,N-K,\alpha}\right)$, where δ is defined in (5.6).

A few important properties of this power function include:

- The power is an increasing function in δ
- The power depends on the true values of the treatment means μ_i, the within group error variance σ^2, and sample size n_i.
- The degrees of freedom used in the power function are $k - 1$ for the numerator degrees of freedom (k is the number of groups) and $N - k$ for the denominator degrees of freedom, where $N = \sum_{i=1}^{k} n_i$ (n_i is the total number of observations in group i) is the total number of observations.

An investigator could also design a study by specifying an effect size. In this case, the effect size is the standard deviation of the population means divided by the common within-population standard deviation.

$$f = \sqrt{\frac{\sum_{i=1}^{k} \left(\mu_i - \bar{\mu}\right)^2 /k}{\sigma^2}}.$$

$\bar{\mu} = \sum_{i=1}^{k} \mu_i/k$, and σ^2 is the within group error variance.

When the number of experimental units per group is constant, the relationship between effect size f for ANOVA and the non-centrality parameter δ is

$$\delta = knf^2, \tag{5.7}$$

where $n_i = n$ is the number of observations in group $i = 1, ..., k$.

5.7.1   Computation Lab: Sample Size for ANOVA

5.7.1.1   Power Function

Power of the ANOVA test can be computed directly by writing a function to compute (5.6) and then using the R functions qf() and pf() for the F quantile and distribution functions.

```
anovancp <- function(n, mu, sigma) {
  mubar <- mean(mu)
  sum(n * (mu - mubar) ^ 2) / sigma ^ 2
}
```

Suppose that an investigator hypothesizes that the true means are $\mu_1 = 1$, $\mu_2 = 1.5$, $\mu_3 = 0.9$, and $\mu_4 = 1.9$, within group variance $\sigma^2 = 1$, and $n_i = 20$. The power $P\left(F_{3,17}(12.95) > 2.725\right)$ is computed below.

```
n <- rep(20, 4)
ncp <- anovancp(n = n,
                mu = c(1, 1.5, 0.9, 1.9),
                sigma = 1.0)
N <- sum(n)
k <- length(n)

1 - pf(
  q = qf(.95, k - 1, N - k),
  df1 = k - 1,
  df2 = N - k,
  ncp = ncp
)
```

```
## [1] 0.8499
```

Instead, if an investigator wanted to specify an effect size $f = 0.4$, we could use (5.7) to calculate the non-centrality parameter to calculate power.

```
ncp <- (0.4 ^ 2) * 20 * 4
1 - pf(
  q = qf(.95, k - 1, N - k),
  df1 = k - 1,
```

```
  df2 = N - k,
  ncp = ncp
)
```

[1] 0.8454

Alternatively, we can use the function **power.anova.test()**. The advantage of this function is that it can compute power or determine parameters to obtain target power. The same computation as above can be obtained by

```
power.anova.test(
  groups = 4,
  n = 20,
  between.var = var(c(1, 1.5, 0.9, 1.9)),
  within.var = 1
)
```

But, if the investigators would like the sample size per group that corresponds to 80% power, then we can remove **n** and add **power = 0.80**.

```
power.anova.test(
  groups = 4,
  power = 0.8,
  between.var = var(c(1, 1.5, 0.9, 1.9)),
  within.var = 1
)
```

```
##
##       Balanced one-way analysis of variance power calculation
##
##            groups = 4
##                 n = 17.85
##       between.var = 0.2158
##        within.var = 1
##         sig.level = 0.05
##             power = 0.8
##
## NOTE: n is number in each group
```

In this case, we see that 18 experimental units per group will yield a test with 80% power.

5.7.1.2 Simulation

Similiar to Section 4.7 simulation can be used to calculate power.

1. Use the underlying model to generate random data with (a) specified sample sizes, (b) parameter values that one is trying to detect with the hypothesis test, and (c) nuisance parameters such as variances.

2. Run the estimation program (e.g., `anova()`) on these randomly generated data.

3. Calculate the test statistic and p-value.

4. Do Steps 1–3 many times, say, N, and save the p-values. The estimated power for a level alpha test is the proportion of observations (out of N) for which the p-value is less than alpha.

One of the advantages of calculating power via simulation is that we can investigate what happens to power if, say, some of the assumptions behind one-way ANOVA are violated.

An R function that implements 1-4 above is given below.

```
powsim_anova <- function(nsim, mu, sigma, n, alpha) {
  res <- numeric(nsim) # store p-values in res
  for (i in 1:nsim) {
    # generate a random samples
    y1 <- rnorm(n = n[1],
                mean = mu[1],
                sd = sigma[1])
    y2 <- rnorm(n = n[2],
                mean = mu[2],
                sd = sigma[2])
    y3 <- rnorm(n = n[3],
                mean = mu[3],
                sd = sigma[3])
    y4 <- rnorm(n = n[4],
                mean = mu[4],
                sd = sigma[4])

    y <- c(y1, y2, y3, y4)
```

```
  # generate the treatment assignment for each group
  trt <-
    as.factor(c(rep(1, n[1]), rep(2, n[2]),
                 rep(3, n[3]), rep(4, n[4])))
  m <- lm(y ~ trt) # fit model
  res[i] <- anova(m)[1, 5] # p-value of F test
}
return(sum(res <= alpha) / nsim)
}
```

```
nsim <- 1000
mu <- c(1, 1.5, 0.9, 1.9)
sigma <- rep(1, 4)
n <- rep(18, 4)
alpha <- 0.05

powsim_anova(
  nsim = nsim,
  mu = mu,
  sigma = sigma,
  n = n,
  alpha = alpha
)
```

```
## [1] 0.813
```

We can investigate the effect of non-equal within group variance.

```
sigma <- c(1, 1.5, 1, 1.2)
powsim_anova(
  nsim = nsim,
  mu = mu,
  sigma = sigma,
  n = n,
  alpha = alpha
)
```

```
## [1] 0.64
```

In this case, we see that the power decreases by 21%.

5.8 Randomized Block Designs

We have already encountered block designs in randomized paired designs in Section 3.10. In these designs, the block size is two. In general, block designs the size of a block can be larger.

Where do block designs fit into what we have learned so far?

Comparison of two treatments

- Unblocked arrangements: unpaired comparison of two treatment groups
- Blocked arrangements: paired comparison of two treatments

Comparison of more than two treatments

- Unblocked arrangements: randomized one-way design
- Blocked arrangements: randomized block design

In blocked designs, two kinds of effects are contemplated:

1. treatments (this is what the investigator is interested in).

2. blocks (this is what the investigator wants to eliminate due to the contribution to the treatment effect).

Blocks might be different litters of animals; blends of chemical material; strips of land; or contiguous periods of time.

5.8.1 ANOVA Identity for Randomized Block Designs

Let $y_{ij}, i = 1, \ldots, b; j = 1, \ldots, k$ be the measurement for the unit assigned to the jth treatment in the ith block. The total sum of squares can be re-expressed by adding and subtracting the treatment and block averages as:

$$\sum_{i=1}^{b} \sum_{j=1}^{k} \left(y_{ij} - \bar{y}_{..}\right)^2 = \sum_{i=1}^{b} \sum_{j=1}^{k} \left[(\bar{y}_{i\cdot} - \bar{y}_{..}) + (\bar{y}_{\cdot j} - \bar{y}_{..}) + (y_{ij} - \bar{y}_{i\cdot} - \bar{y}_{\cdot j} + \bar{y}_{..})\right]^2 . \quad (5.8)$$

After expanding and simplifying the equation above, it can be shown that:

$$\underbrace{\sum_{i=1}^{b}\sum_{j=1}^{k}\left(y_{ij}-\bar{y}_{..}\right)^2}_{SS_T} = \underbrace{b\sum_{i=1}^{k}\left(\bar{y}_{.j}-\bar{y}_{..}\right)^2}_{SS_{Treat}} + \underbrace{k\sum_{j=1}^{b}\left(\bar{y}_{i.}-\bar{y}_{..}\right)^2}_{SS_{Blocks}} + \underbrace{\sum_{i=1}^{b}\sum_{j=1}^{k}\left(y_{ij}-\bar{y}_{i.}-\bar{y}_{.j}+\bar{y}_{..}\right)^2}_{SS_E}$$

$$(5.9)$$

If the experiment was not blocked then SS_E would include SS_{Blocks}, which means the sum of squares due to error and mean squared error would be larger. In other words, *blocking reduces the variability not due to treatment* among units within each treatment group.

There are $N = nk$ observations, so SS_T has $N-1$ degrees of freedom. SS_{Treat} and SS_{Blocks} have $k-1$ and $b-1$ degrees of freedom, respectively, since there are k treatments and b blocks. The sum of squares on the left-hand side the equation should add to the sum of squares on the right-hand side of the equation. Therefore, the error sum of squares has $(N-1)-(k-1)-(b-1) = (kb-1)-(k-1)-(b-1) = (k-1)(b-1)$ degrees of freedom.

Example 5.3. Randomized block designs were first discussed by Fisher [1937]. The "practical example" he gave was an experiment carried out in the state of Minnesota in 1930 and 1931 reported by Immer, Hayes, and Le Roy [Wright, 2013]. The experiment compared the yield of five different varities of barley. The blocks in this example are twelve separate experiments carried out at six locations in the state. The data are shown in Table 5.4.

TABLE 5.4: Minnesota Barley Data

Block	Manchuria	Svansota	Velvet	Trebi	Peatland
1	81.0	105.4	119.7	109.7	98.3
2	80.7	82.3	80.4	87.2	84.2
3	146.6	142.0	150.7	191.5	145.7
4	100.4	115.5	112.2	147.7	108.1
5	82.3	77.3	78.4	131.3	89.6
6	103.1	105.1	116.5	139.9	129.6
7	119.8	121.4	124.0	140.8	124.8
8	98.9	61.9	96.2	125.5	75.7
9	98.9	89.0	69.1	89.3	104.1
10	66.4	49.9	96.7	61.9	80.3
11	86.9	77.1	78.9	101.8	96.0
12	67.7	66.7	67.4	91.8	94.1

5.8.2   Computation Lab: ANOVA for Randomized Block Design

The data collected for the randomized block design described in Example 5.3 is available in the data frame `fisher.barley` in `agridat::fisher.barley` [Wright, 2013].

```
head(fisher.barley)
```

```
##   yield        gen                env year
## 1  81.0 Manchuria UniversityFarm 1931
## 2  80.7 Manchuria UniversityFarm 1932
## 3 146.6 Manchuria         Waseca 1931
## 4 100.4 Manchuria         Waseca 1932
## 5  82.3 Manchuria         Morris 1931
## 6 103.1 Manchuria         Morris 1932
```

The blocks are a combination of the experiments' locations (`env`) and years (`year`). For example, the first block is (`env == UniversityFarm`) & (`year == 1931`).

```
fisher.barley %>%
  group_by(env, year) %>%
  count()
```

```
## # A tibble: 12 x 3
## # Groups:    env, year [12]
##      env            year      n
##      <fct>         <int> <int>
##  1 Crookston      1931      5
##  2 Crookston      1932      5
##  3 Duluth         1931      5
##  4 Duluth         1932      5
##  5 GrandRapids    1931      5
##  6 GrandRapids    1932      5
##  7 Morris         1931      5
##  8 Morris         1932      5
##  9 UniversityFarm 1931      5
## 10 UniversityFarm 1932      5
## 11 Waseca         1931      5
## 12 Waseca         1932      5
```

A block variable can be created by grouping `fisher.barley` and extracting the `dplyr::group_keys()`, which has one row for each group. The R code below groups the data by `env` and `year`, extracts the `group_keys()`, and creates a new variable called `Block` that is equal to `row_number()`.

```
blocks <- fisher.barley %>%
  group_by(env, year) %>%
  group_keys() %>%
  mutate(Block = as.factor(row_number()))

head(blocks, n = 4)
```

```
## # A tibble: 4 x 3
##    env          year Block
##    <fct>       <int> <fct>
## 1 Crookston    1931 1
## 2 Crookston    1932 2
## 3 Duluth       1931 3
## 4 Duluth       1932 4
```

We can add `Block` to `fisher.barley` using `dplyr::left_join()`.

```
fisher.barley2 <-
  fisher.barley %>%
  left_join(blocks, by = c("env", "year"))

glimpse(fisher.barley2)
```

```
## Rows: 60
## Columns: 5
## $ yield <dbl> 81.0, 80.7, 146.6, 100.4, 82.3, 103.1, ~
## $ gen   <fct> Manchuria, Manchuria, Manchuria, Manchu~
## $ env   <fct> UniversityFarm, UniversityFarm, Waseca,~
## $ year  <int> 1931, 1932, 1931, 1932, 1931, 1932, 193~
## $ Block <fct> 9, 10, 11, 12, 7, 8, 1, 2, 5, 6, 3, 4, ~
```

Applying (5.9) to `fisher.barley2`, the sum of squares due to blocks and treatment can be directly calculated.

The grand average is

```
grandmean <-
  fisher.barley2 %>%
  summarise(yield_ave = mean(yield))
```

The block averages, and the sum of squares of block deviations are:

```
df_block <- fisher.barley2 %>%
  group_by(Block) %>%
  summarise(n = n(), block_ave = mean(yield))

sum((df_block$block_ave - grandmean$yield_ave) ^ 2) *
  nlevels(fisher.barley2$gen)
```

```
## [1] 31913
```

The treatment averages, treatment deviations from the grand average, and the sum of squares of treatment deviations are:

```
df_treat <- fisher.barley2 %>%
  group_by(gen) %>%
  summarise(n = n(), treat_ave = mean(yield))

sum((df_treat$treat_ave - grandmean$yield_ave) ^ 2) *
  nlevels(as.factor(fisher.barley2$Block))
```

```
## [1] 5310
```

The aov() function computes an ANOVA table and the summary() function computes the F value and p-values Pr(>F).

```
barley_aov <-
  aov(yield ~ gen + as.factor(Block), data = fisher.barley2)
summary(barley_aov)
```

```
##                 Df Sum Sq Mean Sq F value  Pr(>F)
```

```
## gen                    4    5310     1327     7.78 7.9e-05 ***
## as.factor(Block)      11   31913     2901    17.00 2.1e-12 ***
## Residuals             44    7509      171
## ---
## Signif. codes:
## 0 '***' 0.001 '**' 0.01 '*' 0.05 '.' 0.1 ' ' 1
```

```
summary(aov(yield ~ gen + as.factor(Block), data = fisher.barley2))
```

```
##                    Df Sum Sq Mean Sq F value  Pr(>F)
## gen                 4    5310    1327    7.78 7.9e-05 ***
## as.factor(Block)   11   31913    2901   17.00 2.1e-12 ***
## Residuals          44    7509     171
## ---
## Signif. codes:
## 0 '***' 0.001 '**' 0.01 '*' 0.05 '.' 0.1 ' ' 1
```

5.9 The Linear Model for Randomized Block Design

The linear model for a randomized block design is

$$y_{ij} = \mu + \beta_i + \tau_j + \epsilon_{ij}, \tag{5.10}$$

where y_{ij} is the jth treatment in the ith block, β_i is the ith block effect, and τ_j is the jth treatment effect. μ represents the overall mean, α_i and β_i stand for row and column effects. α_i tells us how much better or worse the mean of the ith row is than the overall mean, so without restricting the scope of the model we assume that $\sum_i \alpha_i = 0$ and $\sum_j \beta_j = 0$.

Equation (5.10) implies that row and column effects are additive. Ignoring experimental errors, ϵ_{ij}, the difference in effect between column 2 and column 1 in row (block) i is $(\mu + \beta_i + \tau_2) - (\mu + \beta_i + \tau_1) = \tau_2 - \tau_1$. The column difference is the same for all (blocks) rows. This assumption may or may not hold for a particular data set. In other words, this model assumes that there is no interaction between blocks and treatments. An interaction could occur in Example 5.3 if, for example, **Manchuria** variety seeds at University Farm in 1931 were damaged or diseased and therefore ineffective, even though they did not affect the other barley varieties. Another way in which an interaction can occur is when the response relationship is multiplicative

$$E(y_{ij}) = \mu \tau_i \beta_j.$$

Taking logs and denoting transformed terms by primes, the model then becomes

$$y'_{ij} = \mu' + \tau'_i + \beta'_j + \epsilon'_{ij}$$

and assuming that ϵ'_{ij} were approximately independent and identically distributed the response $y'_{ij} = log(y_{ij})$ could be analyzed using a linear model in which the interaction would disappear.

Interactions often belong to two categories: (a) transformable interactions, which are eliminated by transformation of the original data, and (b) nontransformable interactions such as a treatment blend interaction that cannot be eliminated via a transformation.

The residuals ϵ_{ij} are assumed to be i.i.d. $N\left(0, \sigma^2\right)$, and represent how much a data set departs from a strictly additive model.

Least squares estimates $\hat{\mu}$, $\hat{\beta}_i$, and $\hat{\tau}_j$ that minimize the sum of squares of the residuals,

$$\sum_{i=1}^{b} \sum_{j=1}^{k} \left(y_{ij} - \hat{\mu} - \hat{\beta}_i - \hat{\tau}_j\right)^2,$$

are $\hat{\mu} = \bar{y}_{..}$, $\hat{\tau}_j = \bar{y}_{.j} - \bar{y}_{..}$, and $\hat{\beta}_i = \bar{y}_{i.} - \bar{y}_{..}$. The jth treatment effect is the jth treatment mean minus the overall mean.

The normality of the residuals can be assessed by, for example, using a normal quantile plot. The constant variance assumption can be investigated by plotting the predicted values versus the residuals. The predicted (fitted) values are

$$\hat{y}_{ij} = \bar{y}_{i.} + \bar{y}_{.j} - \bar{y}_{..}$$

and the residuals are

$$y_{ij} - \hat{y}_{ij}.$$

5.9.1 Computation Lab: Regression Model for Randomized Block Design

The ANOVA table, treatment, and block effects can be computed by fitting a least squares model using lm() and extracting these elements. A regression model to fit the data from the randomized block design of Example 5.3 can be computed using lm() with the treatment variable **gen** and block variable Block.

```
blockmod <- lm(yield ~ gen + Block, data = fisher.barley2)
broom::tidy(blockmod)
```

```
## # A tibble: 16 x 5
##      term          estimate std.error statistic  p.value
##      <chr>            <dbl>     <dbl>     <dbl>    <dbl>
##  1 (Intercept)      119.       6.75     17.7    1.22e-21
##  2 genPeatland        8.15     5.33      1.53   1.34e- 1
##  3 genSvansota       -3.26     5.33     -0.611  5.44e- 1
##  4 genTrebi          23.8      5.33      4.46   5.53e- 5
##  5 genVelvet          4.79     5.33      0.898  3.74e- 1
##  6 Block2           -34.5      8.26     -4.18   1.37e- 4
##  7 Block3           -38.0      8.26     -4.60   3.55e- 5
##  8 Block4           -48.6      8.26     -5.88   4.99e- 7
##  9 Block5           -36.1      8.26     -4.37   7.55e- 5
## 10 Block6           -55.1      8.26     -6.67   3.47e- 8
## 11 Block7           -34.4      8.26     -4.16   1.45e- 4
## 12 Block8            -7.32     8.26     -0.886  3.80e- 1
## 13 Block9           -23.3      8.26     -2.82   7.08e- 3
## 14 Block10          -43.2      8.26     -5.23   4.52e- 6
## 15 Block11           29.1      8.26      3.53   9.97e- 4
## 16 Block12           -9.38     8.26     -1.14   2.62e- 1
```

How can we interpret the regression coefficients? As a first step we need to know what contrasts are used for the treatment and the blocks.

```
contrasts(fisher.barley2$gen)
```

```
##           Peatland Svansota Trebi Velvet
## Manchuria        0        0     0      0
## Peatland         1        0     0      0
## Svansota         0        1     0      0
## Trebi            0        0     1      0
## Velvet           0        0     0      1
```

```
contrasts(fisher.barley2$Block)
```

```
##   2 3 4 5 6 7 8 9 10 11 12
## 1 0 0 0 0 0 0 0 0  0  0  0
```

```
## 2   1 0 0 0 0 0 0 0   0   0   0
## 3   0 1 0 0 0 0 0 0   0   0   0
## 4   0 0 1 0 0 0 0 0   0   0   0
## 5   0 0 0 1 0 0 0 0   0   0   0
## 6   0 0 0 0 1 0 0 0   0   0   0
## 7   0 0 0 0 0 1 0 0   0   0   0
## 8   0 0 0 0 0 0 1 0   0   0   0
## 9   0 0 0 0 0 0 0 1   0   0   0
## 10  0 0 0 0 0 0 0 0   1   0   0
## 11  0 0 0 0 0 0 0 0   0   1   0
## 12  0 0 0 0 0 0 0 0   0   0   1
```

`fisher.barley2$gen` and `fisher.barley2$Block` use dummy coding.

The regression model with covariates for Example 5.3 is

$$\mu_{ij} = E(y_{ij}) = \mu + \sum_{i=1}^{11} \beta_i B_i + \sum_{j=1}^{4} \tau_j T_j,$$

where,

$$B_i = \begin{cases} 1 & \text{if Block } i = 1, \ldots, 11 \\ 0 & \text{Otherwise.} \end{cases} \quad , T_j = \begin{cases} 1 & \text{if Treatment } i = 1, \ldots, 4 \\ 0 & \text{Otherwise.} \end{cases}$$

Consider, an estimate of treatment $j = 1$ from the regression equation,

$$\mu_{11} = \mu + \beta_1 + \tau_1$$
$$\mu_{21} = \mu + \beta_2 + \tau_1$$
$$\vdots$$
$$\mu_{11\,1} = \mu + \beta_{11} + \tau_1$$
$$\mu_{12\,1} = \mu + \tau_1$$

It follows that $\mu_{.1} = \sum_{j=1}^{12} \mu_{1j}/12 = (12\mu + \sum_{i=1}^{11} \beta_i + 12\tau_1)/12 \Rightarrow \tau_1 = \mu_{.1} - \mu - \sum_{i=1}^{11} \beta_i/12$, and $\mu_{12,5} = \mu$.

We can use the data set and the estimated regression coefficients to verify the treatment effect of Peatland.

$\hat{\tau}_1 = 8.15$ is computed from the data and estimated regression coefficients.

$\mu_{.1}$ is the treatment mean for Peatland.

```
mu1 <-
    mean(fisher.barley2[fisher.barley2$gen == "Peatland",]$yield)
```

$\hat{\mu}$ is Manchuria in block 1.

```
mu <-
  mean(fisher.barley2[fisher.barley2$gen == "Manchuria" &
                     fisher.barley2$Block == "1", ]$yield)
```

$\sum_{i=1}^{11} \hat{\beta}/12$ is

```
beta_ave <- sum(coefficients(blockmod)[6:16]) / 12
```

```
mu1 - mu - beta_ave
```

```
## [1] 7.812
```

The difference between this value and the regression coefficient is due to rounding.

If deviation coding is used for treatments and blocks instead, then

$$B_i = \begin{cases} 1 & \text{if Block } i \\ -1 & \text{if Block } b \\ 0 & \text{Otherwise} \end{cases}, T_j = \begin{cases} 1 & \text{if Treatment } j \\ -1 & \text{if Treatment } k \\ 0 & \text{Otherwise} \end{cases}$$

```
contrasts(fisher.barley2$gen) <- contr.sum(5)
contrasts(fisher.barley2$Block) <- contr.sum(12)
```

```
blockmod <- lm(yield ~ gen + Block, data = fisher.barley2)
tidy(blockmod)
```

```
## # A tibble: 16 x 5
##     term         estimate std.error statistic  p.value
##     <chr>           <dbl>     <dbl>     <dbl>    <dbl>
## 1 (Intercept)      101.       1.69      59.9  7.95e-44
## 2 gen1            -6.70       3.37      -1.99  5.33e- 2
## 3 gen2             1.45       3.37       0.430 6.69e- 1
## 4 gen3            -9.96       3.37      -2.95  5.05e- 3
## 5 gen4            17.1        3.37       5.07  7.58e- 6
```

```
##  6 Block1         25.1        5.59      4.48  5.22e- 5
##  7 Block2        -9.45        5.59     -1.69  9.82e- 2
##  8 Block3        -12.9        5.59     -2.32  2.53e- 2
##  9 Block4        -23.6        5.59     -4.21  1.24e- 4
## 10 Block5        -11.0        5.59     -1.97  5.53e- 2
## 11 Block6        -30.1        5.59     -5.37  2.80e- 6
## 12 Block7        -9.31        5.59     -1.66  1.03e- 1
## 13 Block8         17.7        5.59      3.17  2.75e- 3
## 14 Block9         1.73        5.59     0.309  7.59e- 1
## 15 Block10       -18.1        5.59     -3.24  2.27e- 3
## 16 Block11        54.2        5.59      9.69  1.74e-12
```

Let $\mu_{\cdot j} = \sum_{i=1}^{12} \mu_{ij}/12$, $\mu_{i\cdot} = \sum_{j=1}^{5} \mu_{ij}/5$, and $\mu_{\cdot\cdot} = \sum_{i=1}^{12} \sum_{j=1}^{5} \mu_{ij}/(12 \cdot 5)$. Then we can express the regression coefficients as:

$$\mu = \mu_{\cdot\cdot}$$
$$\tau_j = \mu_{\cdot j} - \mu_{\cdot\cdot}, j = 1, \dots, 4$$
$$\beta_i = \mu_{i\cdot} - \mu_{\cdot\cdot}, i = 1, \dots, 11$$
$$-\sum_{j=1}^{4} \tau_j = \mu_{\cdot 5} - \mu_{\cdot\cdot}$$
$$-\sum_{i=1}^{11} \beta_i = \mu_{12\cdot} - \mu_{\cdot\cdot}$$

5.9.2 Computation Lab: Balanced Randomization via Permuted Block Randomization

One application of blocking is to use it as a tool to achieve a balanced randomization.

Randomizing subjects to, say, two treatments in the design of a clinical trial should produce two treatment groups where all the covariates are balanced. But it doesn't guarantee that equal numbers of patients will be assigned to each treatment group.

Simple randomization assigns subjects to two treatments with probability 1/2. This may cause imbalance among different groups (e.g., male/female).

A permuted block randomization is a way to use blocking to assign experimental units to treatments in the blocks. Suppose we want to compare two treatments, A, B, in a randomized study. In a permuted block randomization we could use blocks of size 4. The possible sequences are:

AABB, ABAB, ABBA, BBAA, BABA, and *BAAB.*

When a subject is to be randomized, a block is randomly chosen. For example, if ABAB is chosen, then the first patient receives A, the second patient receives B, etc. This ensures an equal number of patients in each treatment group.

How can this be done using R?

1. Define a vector `trts` that stores the two treatments.
2. Replicate `trts` twice to form a `block` of two As and two Bs.
3. Randomly shuffle the `block` using `sample`.

```
trts <- c("A", "B")   #1
block <- rep(trts, 2) #2
sample(block) #3
```

```
## [1] "B" "B" "A" "A"
```

5.10 Latin Square Designs

The Latin Square design is another type of design that utilizes the blocking principle. If there is more than one nuisance source (i.e., blocking variable) that can be eliminated in the design, then a Latin Square design might be appropriate. The Latin Square design is appropriate when there are two factors used for blocking that both have the same number of levels as the treatment variables.

In general, a **Latin Square** for p factors, or a $p \times p$ Latin Square, is a table containing p rows and p columns. This is often referred to as a square table since the numbers of rows and columns are the same. Each of the p^2 cells contains one of the p letters that correspond to a treatment. Each letter occurs once and only once in each row and column. There are many possible $p \times p$ Latin squares. Tables 5.5, and 5.6 show examples of Latin Squares for $p = 3$ and $p = 4$. For example, in Table 5.5, Col1, Col2, and Col3 are three levels of a blocking variable and Row1, Row2, and Row3 are three levels of another blocking variable.

Example 5.4. Thailand has one of the highest rates of sodium consumption, with fish sauce being one of the main sources. Kanchanachitra et al. [2020] designed a study to examine whether changes in the microenvironment factors can affect fish sauce consumption behavior in a university setting in Thailand. The primary outcome of interest was fish sauce used (grams)/bowls sold per

TABLE 5.5: 3×3 Latin Square Example

	Col1	Col2	Col3
Row1	B	A	C
Row2	A	C	B
Row3	C	B	A

TABLE 5.6: 4×4 Latin Square Example

	Col1	Col2	Col3	Col4
Row 1	B	A	D	C
Row 2	C	D	A	B
Row 3	D	B	C	A
Row 4	A	C	B	D

TABLE 5.7: Latin Square Design from Kanchanachitra et al. [2020]

	Weeks				
Canteen	1	2	3	4	5
I	A	B	C	E	D
II	B	A	D	C	E
III	E	B	A	D	C
IV	C	E	B	A	D
V	D	C	E	B	A

day. Five interventions including a control intervention were examined in five Thai university restuarants (canteens) over five weeks. Each canteen received one of the interventions (treatments) each week.

Table 5.7 shows the design. Each treatment appears once in every row (Canteen) and column (Weeks). Randomization can be achieved by randomly allocating interventions to the symbols A, B, C, D, and E; canteens to the symbols I, II, III, IV, and V; and weeks to the symbols 1, 2, 3, 4, and 5. The treatments are denoted by Latin letters A, B, C, D, and E; hence the name Latin Square design. The design allows for blocking with two variables. The two blocking variables must have the same number of levels as the treatment variable.

The design in Example 5.4 could have used a single canteen in one week with 25 experimental runs for the five treatments. This design could also be valid, but the Latin Square has the advantage that the results do not just apply to one canteen and one week.

5.10.1   Computation Lab: Randomizing Treatments to a Latin Square

How can t treatments be assigned to a Latin Square design using R?

(1) randomly select a Latin Square then
(2) randomly allocate the t treatments to the first t Latin letters.

Suppose an investigator wants to randomly assign six treatments to a Latin Square design. For example, `magic::rlatin(n = 2, size = 4)` will generate a sequence of two Latin Squares stored in a matrix using the numbers 1 through 4. To generate one 4×4 Latin Square, we can use

```
set.seed(25)
magic::rlatin(n = 1, size = 4)
```

```
##      [,1] [,2] [,3] [,4]
## [1,]    3    1    2    4
## [2,]    4    2    1    3
## [3,]    1    4    3    2
## [4,]    2    3    4    1
```

The next step is to randomly allocate treatments A through D to the numbers 1 through 4. To do this we can take a random permutation of the letters A through D (`LETTERS[1:4]`) and then replace 1 in the Latin Square with the first of the shuffled letters, replace 2 with the second of the shuffled letters, etc.

```
set.seed(25)

numtrts <- 4

# Select a random Latin Square
latsq_design <- magic::rlatin(n = 1, size = numtrts)

# Assign treatments to Latin Letters
trts <- LETTERS[1:numtrts]

# Random permutation of Treatments
rand_trts <- sample(trts)

# Matrix store results
```

```r
latsq_design1 <- matrix(nrow = numtrts, ncol = numtrts)

# Loop through Latin Square to replace numbers
# with shuffled treatments
for (i in 1:numtrts) {
  for (j in 1:numtrts) {
    if (latsq_design[i, j] == 1)
      (latsq_design1[i, j] <- rand_trts[1])
    else if (latsq_design[i, j] == 2)
      (latsq_design1[i, j] <- rand_trts[2])
    else if (latsq_design[i, j] == 3)
      (latsq_design1[i, j] <- rand_trts[3])
    else
      (latsq_design1[i, j] <- rand_trts[4])
  }
}
```

The shuffled treatments are

```r
rand_trts
```

```
## [1] "C" "A" "D" "B"
```

So 1 is replaced with C, 2 is replaced with C, etc. The result is

```r
latsq_design1
```

```
##        [,1] [,2] [,3] [,4]
## [1,] "D"  "C"  "A"  "B"
## [2,] "B"  "A"  "C"  "D"
## [3,] "C"  "B"  "D"  "A"
## [4,] "A"  "D"  "B"  "C"
```

5.11   Statistical Analysis of Latin Square Design

The ANOVA identity for a Latin Square design is similar to the ANOVA identity for a Randomized Block design. The total sum of squares in a Latin Square design can be decomposed into sums of squares components related to treatment, row, and column blocking variables similar to (5.8).

The total sum of squares can be re-expressed by adding and subtracting the treatment and block averages for rows and columns as:

$$SS_T = SS_{Treat} + SS_{\text{Row}} + SS_{\text{Col}} + SS_E.$$

SS_T are the sum of squared deviations of each observation from the grand average, SS_{Treat} is the sum of squared deviations of the treatment means from the grand average, etc.

By a reasoning similar to randomized blocks, the associated degrees of freedom are also additive. The total degrees of freedom are $p^2 - 1$, the treatment, row and column degrees of freedom are each $p - 1$, so the error or residual degrees of freedom are $(p^2 - 1) - 3(p - 1) = (p - 1)(p - 2)$.

The ANOVA table for a Latin Square design is

TABLE 5.8: ANOVA Table for Latin Square

Source of variation	Degrees of freedom	Sum of squares	Mean square	F
Columns	$p - 1$	SS_{Col}	MS_{Col}	$\frac{MS_{\text{Col}}}{MS_E}$
Rows	$p - 1$	SS_{Row}	MS_{Row}	$\frac{MS_{\text{Row}}}{MS_E}$
Treatments	$p - 1$	SS_{Treat}	MS_{Treat}	$\frac{MS_{Treat}}{MS_E}$
Residuals	$(p - 1)(p - 2)$	SS_E	MS_E	
Total	$p^2 - 1$	SS_T		

The statistical model for a $p \times p$ Latin Square is

$$y_{ijk} = \mu + \alpha_i + \tau_j + \beta_k + \epsilon_{ijk}, \tag{5.11}$$

where $i, j, k = 1, ..., p$, y_{ijk} is the observation in the ith row and kth column, and j is the Latin letter in the (i, j) cell of the Latin square, μ is the overall mean, α_i is the row effect, τ_j is the jth treatment effect, β_k is the kth column effect, and ϵ_{ijk} is the random error. Constraints similar to the randomized block apply to this model.

We will assume that ϵ_{ijk} are i.i.d. $N(0, \sigma^2)$.

The normality assumption together with assuming that the null hypothesis of no differences between between treatments (i.e., $\tau_j = 0$) is true implies that the statistics under the last column in Table 5.8 follow F distributions with numerator degrees of freedom equal to the source of variation and denominator degrees of freedom equal to residual degrees of freedom (i.e., $(p-1)(p-2)$). For example,

$$MS_{Treat}/MS_E \sim F_{(p-1),(p-1)(p-2)}.$$

5.11.1 Computation Lab: Analysis of Latin Square Designs

Bliss and Rose [1940] used a 4×4 Latin Square design to assign four treatments of an extract of a parathyroid labelled A, B, C, and D. The dosages were given to four different dogs at four different times. The response is Mg (milligrams) percent serum calcium. The design is shown in Table 5.9

TABLE 5.9: Latin Square Design and Data from Bliss and Rose [1940]

		Weeks		
Dogs	1	2	3	4
1	C	D	B	A
	13.8	17	16	16
2	B	A	C	D
	15.8	14.3	14.8	15.4
3	D	C	A	B
	15	14.5	14	15
4	A	B	D	C
	14.7	15.4	14.8	14

The data are stored in the data frame BR_LatSq.

The row, column, and treatment means are:

```
BR_LatSq %>%
  group_by(Weeks) %>%
  summarise(Weeks_Ave = mean(value))


## # A tibble: 4 x 2
##    Weeks Weeks_Ave
##    <dbl>     <dbl>
```

```
## 1        1        14.8
## 2        2        15.3
## 3        3        14.9
## 4        4        15.1
```

```
BR_LatSq %>%
  group_by(Dogs) %>%
  summarise(Ave_Dog = mean(value))
```

```
## # A tibble: 4 x 2
##     Dogs Ave_Dog
##    <dbl>   <dbl>
## 1      1    15.7
## 2      2    15.1
## 3      3    14.6
## 4      4    14.7
```

```
BR_LatSq %>%
  group_by(Treat) %>%
  summarise(Treat_Ave = mean(value))
```

```
## # A tibble: 4 x 2
##    Treat Treat_Ave
##    <chr>     <dbl>
## 1 A          14.8
## 2 B          15.6
## 3 C          14.3
## 4 D          15.6
```

The grand average is

```
grand_Ave <- mean(BR_LatSq$value)
grand_Ave
```

```
## [1] 15.03
```

The sum of squares due to weeks is obtained by computing the squared difference between the mean of each week (`Weeks_Ave`) and the grand average (grand_Ave) for each dog. The squared deviations for Week 1 will be the same for all four dogs, the squared deviations for Week 2 will be the same for all four dogs, etc. So, each squared deviation is multiplied by 4, 4*(Weeks_Ave - grand_Ave)^2.

```
BR_LatSq %>% group_by(Weeks) %>%
  summarise(Weeks_Ave = mean(value)) %>%
  mutate(week_dev = 4 * (Weeks_Ave - grand_Ave) ^ 2)
```

```
## # A tibble: 4 x 3
##    Weeks Weeks_Ave week_dev
##    <dbl>     <dbl>    <dbl>
## 1      1      14.8   0.170
## 2      2      15.3   0.289
## 3      3      14.9   0.0689
## 4      4      15.1   0.0189
```

Finally, we can sum week_dev to obtain the sum of squares due to Weeks.

```
BR_LatSq %>% group_by(Weeks) %>%
  summarise(Weeks_Ave = mean(value)) %>%
  mutate(week_dev = 4 * (Weeks_Ave - grand_Ave) ^ 2) %>%
  summarise(sum(week_dev))
```

```
## # A tibble: 1 x 1
##    `sum(week_dev)`
##             <dbl>
## 1           0.547
```

A function to compute the sum of squares for the blocking variables and treatment is given below. Note that the results of summarise() are piped into first() so that the function returns the first item instead of a data frame.

```
computeSS <- function(a) {
  BR_LatSq %>%
    group_by({{a}}) %>%
    summarise(Weeks_Ave = mean(value)) %>%
    mutate(week_dev = 4 * (Weeks_Ave - grand_Ave) ^ 2) %>%
    summarise(sum(week_dev)) %>%
```

```
    first()
}
```

The degrees of freedom for weeks is $4 - 1 = 3$, so the mean square due to Weeks is

```
MS_Weeks <- computeSS(Weeks) / 3
```

The total sum of squares is

```
sum((BR_LatSq$value - grand_Ave) ^ 2)
```

```
## [1] 11.29
```

So, the mean squared error for this data is obtained by computing the error sum of squares (error_SS) and dividing by its degrees of freedom $(4 - 1)(4 - 2) = 6$:

```
error_SS <-
  sum((BR_LatSq$value - grand_Ave) ^ 2) -
  (computeSS(Weeks) + computeSS(Dogs) + computeSS(Treat))
MS_error <- error_SS / 6
```

The ratio of of the mean square for Weeks to the mean square for error can be compared to an $F_{3,6}$ to compute the p-value to test if the means for days are different:

```
pf(MS_Weeks / MS_error, 3, 6, lower.tail = FALSE)
```

```
## [1] 0.7936
```

The statistical model (5.11) for a 4×4 Latin Square can be fit to the data from 5.9 using lm().

```
blissmod <-
  lm(value ~ as.factor(Weeks) +
       as.factor(Dogs) +
       as.factor(Treat),
     data = BR_LatSq)
broom::tidy(blissmod)
```

```
## # A tibble: 10 x 5
##    term                    estimate std.error statistic p.value
##    <chr>                      <dbl>     <dbl>     <dbl>   <dbl>
## 1 (Intercept)                15.2       0.574    26.5    1.90e-7
## 2 as.factor(Weeks)2           0.475     0.513     0.926  3.90e-1
## 3 as.factor(Weeks)3           0.0750    0.513     0.146  8.89e-1
## 4 as.factor(Weeks)4           0.275     0.513     0.536  6.11e-1
## 5 as.factor(Dogs)2           -0.625     0.513    -1.22   2.69e-1
## 6 as.factor(Dogs)3           -1.08      0.513    -2.10   8.10e-2
## 7 as.factor(Dogs)4           -0.975     0.513    -1.90   1.06e-1
## 8 as.factor(Treat)B           0.8       0.513     1.56   1.70e-1
## 9 as.factor(Treat)C          -0.475     0.513    -0.926  3.90e-1
## 10 as.factor(Treat)D          0.8       0.513     1.56   1.70e-1
```

The treatment effects of B compared to A (the reference category) and C compared to A are: 0.8, and -0.475. The `Residual standard error` is the square-root of the mean square error, `sqrt(error_SS/6)`.

A normal Q-Q plot of the residuals shows that there may be a systematic pattern, although it's difficult assess with 16 observations.

```
tibble(y = blissmod$residuals) %>%
  ggplot(aes(sample = y)) +
  geom_qq() +
  geom_qq_line()
```

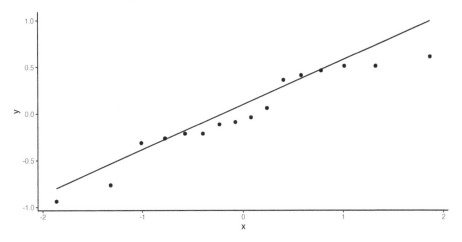

The ANOVA table is obtained for this model using `anova()`.

```
anova(blissmod)
```

```
## Analysis of Variance Table
##
## Response: value
##                    Df Sum Sq Mean Sq F value Pr(>F)
## as.factor(Weeks)    3   0.55   0.182    0.35   0.79
## as.factor(Dogs)     3   2.83   0.944    1.79   0.25
## as.factor(Treat)    3   4.76   1.586    3.01   0.12
## Residuals           6   3.16   0.526
```

The Latin Square model is similar to the Randomized Block design except with an extra blocking factor. Suppose that the design was only blocked on `Dogs`, then the mean square error decreases since `Residuals Df` increases from 6 to 6 + 3 = 9, and `Residuals Sum Sq` increases from 3.16 to 3.16 + 0.55 = 3.71). This has an effect of decreasing the mean square error, so increasing the observed `F value` and decreasing the p-value for `Treat` from 0.12 to 0.05. In this case, accounting for part of the error variance due to `Weeks` could change the conclusion about the effect of treatment.

It assumed that the effects of treatments, weeks, and dogs are all additive so that there are no interaction effects. It can be misleading to a use Latin Square design to study factors that can interact, and can lead to mixing up the effects of one factor with interactions. Outliers can occur as a result of these interactions.

```
blissmod1 <-
  lm(value ~ as.factor(Dogs) + as.factor(Treat), data = BR_LatSq)
anova(blissmod1)
```

```
## Analysis of Variance Table
##
## Response: value
##                  Df Sum Sq Mean Sq F value Pr(>F)
## as.factor(Dogs)   3   2.83   0.944    2.29   0.15
## as.factor(Treat)  3   4.76   1.586    3.85   0.05 .
## Residuals         9   3.71   0.412
## ---
## Signif. codes:
## 0 '***' 0.001 '**' 0.01 '*' 0.05 '.' 0.1 ' ' 1
```

5.12   Graeco-Latin Square Designs

A Graeco-Latin Square is a $k \times k$ pattern that permits study of k treatments
simultaneously with three different blocking variables, where each has k levels.
This is a Latin Square in which each Greek letter appears once and only once
with each Latin letter. It can be used to control three sources of extraneous
variability (i.e. block in three different directions).

To generate a 3×3 Graeco-Latin Square design, superimpose two Latin Square
designs using the Greek letters for the second 3×3 Latin Square. For example,
the Latin Squares in Tables 5.10 and 5.11 can be superimposed by using Greek
letters α, β, γ in place of A, B, C in Table 5.11, and copying the values into
5.10 we obtain the Graeco-Latin Square design in Table 5.12.

TABLE 5.10: 3×3 Latin Square #1 for Graeco-Latin Square

	Col 1	Col 2	Col 3
Row 1	B	A	C
Row 2	A	C	B
Row 3	C	B	A

Table 5.12 is a Graeco-Latin design with three blocking variables represented
by the levels of Row, Col and the Greek letters α, β, and γ.

TABLE 5.11: 3×3 Latin Square #2 for Graeco-Latin Square

	Latin Letters				Greek Letters		
	Col 1	Col 2	Col 3	$\longrightarrow$	Col 1	Col 2	Col 3
Row 1	A	B	C		Row 1 α	β	γ
Row 2	C	A	B		Row 2 γ	α	β
Row 3	B	C	A		Row 3 β	γ	α

TABLE 5.12: 3×3 Graeco Latin Square

	Col 1	Col 2	Col 3
Row 1	B α	A β	C γ
Row 2	A γ	C α	B β
Row 3	C β	B γ	A α

5.12.1 Computation Lab: Graeco-Latin Square Designs

The `rlatin()` function in the `magic` library can be used to generate $p \times p$ Latin Squares that can be combined into Graeco-Latin Squares. The design generated from `rlatin()` can be converted into an R data frame that can be used for statistical analysis.

For example, combining `ls1` and `ls2` will form a 4×4 Graeco-Latin Square.

```
ls1 <- magic::rlatin(4)
ls2 <- magic::rlatin(4)
```

5.13 Exercises

Exercise 5.1. Use R to create a randomization scheme to randomize 222 subjects to three treatments such that there are an equal number of subjects assigned to each treatment. You may use the functions developed in Section 5.2.1.

Exercise 5.2. Explain when it is appropriate to use a randomized block design.

Exercise 5.3. Show that

$$SS_T = SS_{Treat} + SS_E,$$

where $SS_T = \sum_{i=1}^{k} \sum_{j=1}^{n_i} (y_{ij} - \bar{y}_{..})^2$, $SS_{Treat} = \sum_{i=1}^{k} n_i (\bar{y}_{i.} - \bar{y}_{..})^2$, and $SS_E = \sum_{i=1}^{k} \sum_{j=1}^{n_i} (y_{ij} - \bar{y}_{i.})^2$.

Exercise 5.4. Use the R functions `SST()`, `SSTreat()`, and `SSe()` defined in Section 5.3.3 to answer the following questions.

a. Verify the ANOVA identity (Equation (5.3)) using the functions and `painstudy` data frame from Example 5.1.

b. Explain what the following code does in `SSTreat()`.

```
lapply(split(y, groups), function(x) { (mean(x) - mean(y))^2 })
```

c. How and why is the code from part b. different from the following line used in `SSe()`?

```
lapply(split(y, groups), function(x) { (x - mean(x))^2 })
```

Exercise 5.5. Consider the statistical model defined by Equation (5.1) and suppose $H_0 : \tau_1 = \cdots = \tau_k = 0$ is true. Show that the following are true.

a. $SS_{Treat} / \sigma^2 \sim \chi^2_{k-1}$.

b. $SS_E / \sigma^2 \sim \chi^2_{N-k}$.

c. $MS_{Treat} / MS_E \sim F_{k-1, N-k}$.

Exercise 5.6. Explain why the sample variance formula is

$$\sum_{i=1}^{N} \frac{(y_i - \bar{y})^2}{N - 1}$$

instead of

$$\sum_{i=1}^{N} \frac{(y_i - \bar{y})^2}{N}.$$

Exercise 5.7. The code to produce the ANOVA table for Example 5.1 is shown below. Use `pf()` to compute the p-value for this F test. Verify that you get the same answer.

```
anova_mod <- aov(pain ~ trt, data = painstudy)
summary(anova_mod)
```

Exercise 5.8. The following data are the response times (in minutes) of six people measured after two treatments. The order in which each person received the treatments was determined by randomization. The investigator was interested to see if the two treatments are different.

	1	2	3	4	5	6
Treatment I	46	64	80	71	99	70
Treatment II	78	66	70	64	46	70

a. What is the name of this design? Explain.

b. What is the blocking factor used in this design?

c. Is there any evidence at the 5% level that the two treatments are different?

Exercise 5.9. In Section 5.5, both dummy coding and deviation coding were used to code the treatment variable for the data in Example 5.1. Use contr.helmert() to code the treatment variable using the Helmeert coding and fit the linear model in the example.

a. Use R to estimate the regression coefficients. Interpret the estimates.

b. Compare the interpretability of the estimated coefficients using contr.treatment(), contr.sum(), and contr.helmert(). Which contrasts lead to the most appropriate parameter estimates for this application?

Exercise 5.10. Consider the **painstudy2** data frame from Example 5.2.

a. Use pt() and qt() to compute the p-value and 95% confidence interval for the mean comparison of treatments A and B, A - B without any adjustment.

b. Use pt() and qt() to compute the p-value and 95% confidence interval for the mean comparison of treatments A and B, A - B adjusted using the Bonferroni method.

c. Use ptukey() and qtukey() to compute the p-value and 95% confidence interval for the mean comparison of treatments A and B, A - B using the Tukey method.

d. Repeat parts a. to c. using `emmeans()` and verify that you get the same answers.

Exercise 5.11. In Section 5.7.1.2, we investigated the effect of non-equal within group variances on the power of an experiment. Investigate the effect of non-normality by modifying the simulation.

Exercise 5.12. Let y_{ij} for $i = 1, \ldots, b$ and $j = 1, \ldots, k$ be the measurement for the unit assigned to the jth treatment in the ith block.

a. Show that

$$SS_T = SS_{Treat} + SS_{Blocks} + SS_E,$$

where

$$SS_T = \sum_{i=1}^{b} \sum_{j=1}^{k} \left(y_{ij} - \bar{y}_{..} \right)^2,$$

$$SS_{Treat} = b \cdot \sum_{i=1}^{k} \left(\bar{y}_{.j} - \bar{y}_{..} \right)^2,$$

$$SS_{Blocks} = k \cdot \sum_{j=1}^{b} \left(\bar{y}_{i.} - \bar{y}_{..} \right)^2, \text{ and}$$

$$SS_E = \sum_{i=1}^{b} \sum_{j=1}^{k} \left(y_{ij} - \bar{y}_{i.} - \bar{y}_{.j} + \bar{y}_{..} \right)^2.$$

b. Show that blocking reduces the variability not due to treatment among units within each treatment group.

Exercise 5.13. Consider the `fisher.barley` data frame from Example 5.3.

a. Use R to compute SS_E. What are the degrees of freedom for SS_E?

b. A statistician analyzing `fisher.barley` forgot to account for blocking. Estimate the treatment effects without accounting for blocking and compare these estimates when blocking is taken into account.

c. Compute the ANOVA table for `fisher.barley` with and without blocking. Which columns change? Why?

Exercise 5.14. Consider a randomized block design with k treatments and b blocks.

a. Derive the least squares estimators of the treatment effects using i) dummy coding and ii) deviation coding.

b. Verify the estimators with the `fisher.barley` data frame from Example 5.3.

Exercise 5.15. As the statistician on a multidisciplinary research team, you are asked to create a randomization scheme to assign three treatments A, B, and C, to 150 units in blocks of 6. The scheme must ensure that each treatment is assigned to 50 units and that a block is randomly selected when a unit is randomized.

a. How many permuted blocks are required?

b. Use R to create a randomization scheme that satisfies the requirements. The program should return a data frame where each row represents a unit with three columns representing the assigned block, the assigned treatment, and the unit number from 1 to 150.

Exercise 5.16. Randomly permute each row of the matrix below using R.

$$\begin{bmatrix} 1 & 1 & 1 & 1 \\ 2 & 2 & 2 & 2 \\ 3 & 3 & 3 & 3 \\ 4 & 4 & 4 & 4 \end{bmatrix}$$

You can construct the matrix below with `matrix(1:4, nrow = 4, ncol = 4)` in R. Does this permuted matrix form a Latin square? Explain.

Exercise 5.17. Consider the 4x4 square matrix below.

$$\begin{bmatrix} A & B & C & D \\ D & A & B & C \\ C & D & A & B \\ B & C & D & A \end{bmatrix}$$

Suppose the rows correspond to subjects and that each subject is tested under experimental conditions represented by letters in the particular arrangement of the letters across rows. The columns correspond to time periods. It is known that the study outcome under the B condition preceded by the A condition is always lower.

a. Is the matrix a Latin Square design?

b. If treatments are randomly assigned to A through D, will this result in an unbiased treatment effect? Explain.

Exercise 5.18. Consider the Latin Square design and data from Bliss and Rose [1940] shown in Table 5.9.

a. Use R to compute the ANOVA table. Interpret the results.

b. Estimate the treatment effects. Interpret the results.

6

Factorial Designs at Two Levels—2^k Designs

6.1 Introduction

A factorial design is an experimental design where the investigator selects a fixed number of factors with $k \geq 2$ levels and then measures response variables at each factor-level combination. The factors can be quantitative or qualitative. For example, two levels of a quantitative variable could be two different temperatures or two different concentrations. Qualitative factors might be two types of catalysts or the presence and absence of some entity.

A 2^k factorial design is notation for a factorial design where:

- the number of factors is $k \geq 2$,

- the number of levels of each factor is 2, and

- the total number of experimental conditions, called *runs*, in the design is 2^k.

In general, factorial experiments can involve factors with different numbers of levels. For example, a factorial experiment that involves three factors with three levels and two factors with two levels would be a $3^3 \times 2^2$ design.

Example 6.1. An investigator is interested in examining three components of a weight loss intervention.

A. Keeping a food diary

B. Increasing activity

C. Home visit

The investigator plans to investigate all $2 \times 2 \times 2 = 2^3 = 8$ combinations of experimental conditions.

The experimental conditions are outlined in Table 6.1. The notation $(-, +)$ is often used to represent the two factor levels in a 2^k design. In this case,

– represents No and + represents Yes. The response variable for the ith run ($i = 1, \ldots, 8$) is represented by y_i, and the observed value is recorded in brackets next to y_i.

TABLE 6.1: Weight Loss Experiment Planning Matrix

A	B	C	Weight loss	Run order	Run
No (−)	No (−)	No (−)	y_1 (1.1)	5	1
No (−)	No (−)	Yes (+)	y_2 (1.8)	8	2
No (−)	Yes (+)	No (−)	y_3 (-0.3)	1	3
No (−)	Yes (+)	Yes (+)	y_4 (-1.1)	2	4
Yes (+)	No (−)	No (−)	y_5 (1.0)	4	5
Yes (+)	No (−)	Yes (+)	y_6 (2.6)	7	6
Yes (+)	Yes (+)	No (−)	y_7 (-0.4)	3	7
Yes (+)	Yes (+)	Yes (+)	y_8 (0.4)	6	8

6.2   Factorial Effects

In ANOVA, the objective is to compare the *individual* experimental conditions with each other. In a factorial experiment, the objective is generally to compare *combinations* of experimental conditions.

What is the effect of keeping a food diary in Example 6.1?

6.2.1   Main Effects

We can estimate the effect of a food diary by comparing the mean of all conditions where food diary is set to NO (Run 1-4) and the mean of all conditions where food diary is set to YES (Run 5-8) (see Table 6.1). This is also called the **main effect** of food diary, the adjective *main* being a reminder that this average is taken over the levels of the other factors.

Let $\bar{y}(A+)$ be the average of the y values at the + level of A, and $\bar{y}(A-)$ be the average of the y values at the − levels of A.

The main effect of food diary, $ME(A)$, is

$$ME(A) = \bar{y}(A+) - \bar{y}(A-) = \frac{y_1 + y_2 + y_3 + y_4}{4} - \frac{y_5 + y_6 + y_7 + y_8}{4}$$

$$= \frac{1.1 + 1.8 - 0.3 - 1.1}{4} + \frac{1.0 + 2.6 - 0.4 + 0.4}{4}$$

$$= 0.525.$$

Similarly, the main effects of physical activity (B) and home visit (C) are:

$$ME(B) = \bar{y}(B+) - \bar{y}(B-) = \frac{y_1 + y_2 + y_5 + y_6}{4} - \frac{y_3 + y_4 + y_7 + y_8}{4} = -1.975,$$

$$ME(C) = \bar{y}(C+) - \bar{y}(C-) = \frac{y_1 + y_3 + y_5 + y_7}{4} - \frac{y_2 + y_4 + y_6 + y_8}{4} = 0.575$$

Figure 6.1 shows a main effects plot for Example 6.1. The averages of all observations at each level of the factor are shown as dots connected by lines. The plot shows that keeping a food diary and a home visit increase weight loss, and increased physical activity decreases weight loss.

All experimental subjects are used, but are rearranged to make each comparison. In other words, subjects are *recycled* to measure different effects. This is one reason why factorial experiments are more efficient.

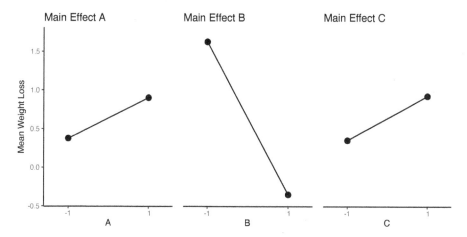

FIGURE 6.1: Main Effects of Example 6.1

6.2.2 Interaction Effects

Let $\bar{y}(A + |B-)$ be the average of y at the $+$ level of A when B is at the $-$ level. The conditional main effect of A at the $+$ level of B is then $ME(A|B+) = \bar{y}(A + |B+) - \bar{y}(A - |B+)$. AB and ABC will be used for notation for the interaction between factors A and B, and A, B, and C.

6.2.2.1   Two Factor Interactions

The effect of food diary when home visit is No is

$$
\begin{aligned}
ME(A|B-) &= \bar{y}(A + |B-) - \bar{y}(A - |B-) \\
&= \frac{(y_5 + y_6)}{2} - \frac{(y_1 + y_2)}{2} \\
&= \frac{(1.0 + 2.6)}{2} - \frac{(1.1 + 1.8)}{2} \\
&= 0.35.
\end{aligned}
$$

The effect of food diary when home visit is Yes is

$$
\begin{aligned}
ME(A|B+) &= \bar{y}(A + |B+) - \bar{y}(A - |B+) \\
&= \frac{(y_7 + y_8)}{2} - \frac{(y_3 + y_4)}{2} \\
&= \frac{((-0.4) + 0.4)}{2} - \frac{(-0.3 + (-1.1))}{2} \\
&= 0.70.
\end{aligned}
$$

The average difference between $ME(A|B-)$ and $ME(A|B+)$ is the *interaction* between factors A and B

$$
\begin{aligned}
INT(A, B) &= \frac{1}{2}[ME(A|B+) - ME(A|B-)] \\
&= \frac{1}{2}[0.70 - 0.35] \\
&= 0.175.
\end{aligned}
$$

If there is a large difference between $ME(A|B-)$ and $ME(A|B+)$, then the effect of factor A may depend on the level of B. In this case, we say that there is an interaction between A and B.

The interaction between A and B from Example 6.1 is shown in Figure 6.2. This plot shows the means of y at each level of A and B. If the distance between ▲ and ● at each level of B is approximately the same, then the main effect of A doesn't depend on the level of B. These plots can often be more informative than the single number of the interaction effect.

6.2.3   Three Factor Interactions

The interaction between food diary (A) and increased physical activity (B) in Example 6.1 when home visit (C) is No is $INT(A, B|C-)$, and $INT(A, B|C+)$ when home visit is Yes. The average difference between $INT(A, B|C-)$ and $INT(A, B|C+)$ is the three-way interaction between A, B, C or $INT(A, B, C)$.

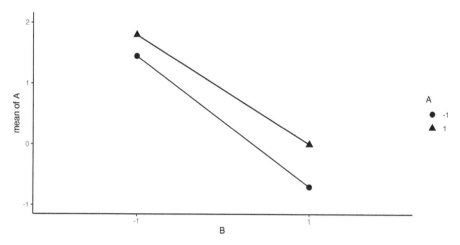

FIGURE 6.2: Interaction between Food Diary and Physical Activity (Example 6.1)

When C = 1 the interaction between A and B is

$$\frac{1}{2}[ME(A|B+,C+) - ME(A|B-,C+)] = \frac{1}{2}[(y_8 - y_4) - (y_6 - y_2)]$$
$$= \frac{1}{2}[(0.4 - (-1.1)) - (2.6 - 1.8)]$$
$$= 0.35.$$

When C = -1 the interaction between A and B is

$$\frac{1}{2}[ME(A|B+,C-) - ME(A|B-,C-)] = \frac{1}{2}[(y_7 - y_3) - (y_5 - y_1)]$$
$$= \frac{1}{2}[(-0.4 - (-0.3)) - (1.0 - 1.1)]$$
$$= 0.$$

Now, we take the average difference to get the three-factor interaction.

$$INT(A, B, C) = \frac{1}{2}[INT(A, B|C+) - INT(A, B|C-)]$$
$$= \frac{1}{2}[0.35 - 0]$$
$$= 0.175.$$

6.2.4   Cube Plot and Model Matrix

Figure 6.3 shows a **cube plot** for weight loss in Example 6.1. The cube plot shows the observations for each factor-level combination in a 2^3 design. The value of y for the various combinations of factors A, B, and C is shown at the corners of the cube. For example, $y = 1.8$ was obtained when A = -1, B = -1, and C = 1. The cube shows how this design produces 12 comparisons along the 12 edges of the cube: four measures of the effect of food diary change; four measures of the effect of increased physical activity; four measures of home visit. On each edge of the cube, only one factor is changed with the other two held constant.

The same information is also displayed in the first three columns of the **model matrix** in Table 6.2. The first column has single -1 and 1 alternating, in the second pairs of -1 and 1 are alternating, etc. When a full factorial design matrix is specified in this manner the design is said to be in *standard order*. The interactions AB, AC, BC, ABC are the products of the columns A and B, A and C, etc.

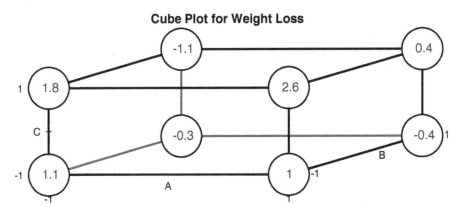

FIGURE 6.3: Cube Plot for Weight Loss from Example 6.1

The model matrix can be be used to calculate factorial effects. For example, the main effect for B can be calculated by multiplying the B column and y column in Table 6.2 and dividing by 4.

6.2.5   Computation Lab: Computing and Plotting Factorial Effects

In this section we will use R to compute factorial effects for the data in Example 6.1. The data are stored in the data frame wtlossdat.

TABLE 6.2: Model Matrix for 2^3 Design and Data from Example 6.1

A	B	C	AB	AC	BC	ABC	y
-1	-1	-1	1	1	1	-1	1.1
-1	-1	1	1	-1	-1	1	1.8
-1	1	-1	-1	1	-1	1	-0.3
-1	1	1	-1	-1	1	-1	-1.1
1	-1	-1	-1	-1	1	1	1.0
1	-1	1	-1	1	-1	-1	2.6
1	1	-1	1	-1	-1	-1	-0.4
1	1	1	1	1	1	1	0.4

(Model Matrix columns: A B C AB AC BC ABC | Data: y)

```
glimpse(wtlossdat, n = 3)
```

```
## Rows: 8
## Columns: 5
## $ A          <dbl> -1, -1, -1, -1, 1, 1, 1, 1
## $ B          <dbl> -1, -1, 1, 1, -1, -1, 1, 1
## $ C          <dbl> -1, 1, -1, 1, -1, 1, -1, 1
## $ y          <dbl> 1.1, 1.8, -0.3, -1.1, 1.0, 2.6, -~
## $ `Run Order` <dbl> 5, 8, 1, 2, 4, 7, 3, 6
```

The main effect of a factor is the mean difference when a factor is at its high versus low level. The main effect of factor A is $\bar{y}(A+) - \bar{y}(A-)$.

```
mean(wtlossdat["y"][wtlossdat["A"] == 1]) -
  mean(wtlossdat["y"][wtlossdat["A"] == -1])
```

```
## [1] 0.525
```

wtlossdat["y"] selects the column y from wtlossdat data frame.

Let's write a function to compute the main effect to avoid copying and pasting similar code for each factor. The function takes a data frame (df), dependent variable (y1), and factor (fct1) that we want to calculate the main effect.

```
calcmaineff <- function(df, y1, fct1) {
  mean(df[y1][df[fct1] == 1]) - mean(df[y1][df[fct1] == -1])
}
```

We can use the function to calculate the main effects of B and C.

```
calcmaineff(df = wtlossdat, y1 = "y", fct1 = "B")
```

```
## [1] -1.975
```

```
calcmaineff(df = wtlossdat, y1 = "y", fct1 = "C")
```

```
## [1] 0.575
```

Another method to calculate factorial effects is to use the design matrix. For example, the main effect of C is the dot product of the C column and the y column divided by 4.

```
sum(wtlossdat["C"] * wtlossdat["y"]) / 4
```

```
## [1] 0.575
```

Figure 6.4 uses ggplot() to plot the main effects.

```
wtlossdat %>%
  group_by(A) %>%
  summarise(mean = mean(y)) %>%
  ggplot(aes(A, mean)) +
  geom_point(size = 3) + geom_line() +
  ggtitle("Main Effect A") +
  ylab("Mean Weight Loss") +
  scale_x_discrete(limits = c(-1, 1))
```

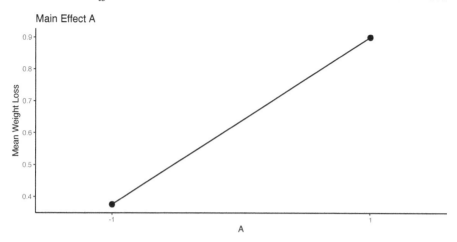

FIGURE 6.4: Main Effect of A Example 6.1

The two factor interaction between A and B is calculated by the following

1. Compute the mean difference between (i) y when `A == 1` and `B == 1`; and (ii) y when `A == -1` and `B == 1`.
2. Compute the mean difference between (i) y when `A == 1` and `B == -1`; and (ii) y when `A == -1` and `B == -1`.
3. Compute the mean difference between 1. and 2.

```
# 1.
abplus <-
  (mean(wtlossdat$y[(wtlossdat$A == 1 & wtlossdat$B == 1)]) -
     mean(wtlossdat$y[(wtlossdat$A == -1 & wtlossdat$B == 1)]))
# 2.
abminus <-
  (mean(wtlossdat$y[(wtlossdat$A == 1 & wtlossdat$B == -1)]) -
     mean(wtlossdat$y[(wtlossdat$A == -1 & wtlossdat$B == -1)]))
# 3.
(abplus - abminus) / 2
```

```
## [1] 0.175
```

We can also modify the function we wrote to compute main effects for two-way interactions.

```
calcInteract2 <- function(df, y1, fct1, fct2) {
  yzplus <- mean(df[y1][df[fct1] == 1 & df[fct2] == 1]) -
    mean(df[y1][df[fct1] == -1 & df[fct2] == 1])

  yzminus <- mean(df[y1][df[fct1] == 1 & df[fct2] == -1]) -
    mean(df[y1][df[fct1] == -1 & df[fct2] == -1])

  return(0.5 * (yzplus - yzminus))
}
```

The three two-way interactions are:

```
calcInteract2(wtlossdat, "y", "A", "B")
```

```
## [1] 0.175
```

```
calcInteract2(wtlossdat, "y", "A", "C")
```

```
## [1] 0.625
```

```
calcInteract2(wtlossdat, "y", "B", "C")
```

```
## [1] -0.575
```

Interactions can also be calculated using the design matrix. For example, the BC interaction is the dot product of columns B, C, and y divided by 4.

```
sum(wtlossdat["B"] * wtlossdat["C"] * wtlossdat["y"]) / 4
```

```
## [1] -0.575
```

Figure 6.5 shows an interaction plot of the AB interaction using interaction.plot().

```
interaction.plot(
  wtlossdat$B,
  wtlossdat$A,
  wtlossdat$y,
  trace.label = "A",
  xlab = "B",
  ylab = "Mean weight loss"
)
```

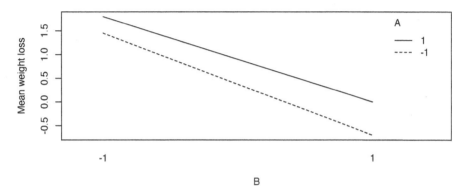

FIGURE 6.5: AB Interaction for Example 6.1

The code below uses **ggplot()** to create an interaction plot similar to Figure 6.5, although there is an extra step where the means are computed and stored in a data frame.

```
wtlossdat %>%
  group_by(A, B) %>%
  summarise(mean = mean(y)) %>%
  ggplot(aes(as.factor(B), mean, group = A)) +
  ylim(-1, 2) +
  geom_line(aes(linetype = as.factor(A))) +
  xlab("B") +
  ylab("mean of A") +
  guides(linetype = guide_legend(title = "A"))
```

Cube plots can be created using **FrF2::cubePlot()**. A simple way to use **cubePlot()** is to input a linear model object of the factorial design (discussed in Section 6.4). The **modeled = FALSE** argument indicates that we want to show the averages instead of the model means. The code below was used to create Figure 6.3.

```
mod <- lm(y ~ A * B * C, data = wtlossdat)
FrF2::cubePlot(
  mod,
  eff1 = 'A',
  eff2 = 'B',
  eff3 = 'C',
  main = "Cube Plot for Weight Loss",
  round = 2,
  modeled = F,
  cex.title = 1
)
```

6.3 Replication in Factorial Designs

Replicating a run is not always feasible in an experiment where each run is costly, but when it can be done, this allows the investigator to estimate variance. Suppose that a 2^3 factorial design is replicated twice. The ith run, $i = 1, \ldots, 8$, has $j = 1, 2$ observations y_{ij}. Assume that $Var(y_{ij}) = \sigma^2$, duplicates within a run are independent, and runs are independent. The data with model matrix is shown in Table 6.3: Rep1 and Rep2 indicate the first and second replications, and Var is the sample variance of a run.

The estimated variance for each experimental run in a duplicated 2^3 design is

$$S_i^2 = \frac{(y_{i1} - y_{i2})^2}{2}.$$

The average of these estimated variances results in a *pooled* estimate of σ^2, $S^2 = \frac{\sum_{i=1}^{8} S_i^2}{8}$.

A 2^k design with m_i replications in the ith run has pooled estimate of σ^2

$$S^2 = \frac{(m_1 - 1)S_1^2 + \cdots + (m_N - 1)S_N^2}{((m_1 - 1) + \cdots + (m_N - 1))}, \tag{6.1}$$

$m_i - 1$ is the degrees of freedom for run i. S^2 has $\sum_{i=1}^{2^k}(m_i - 1)$ degrees of freedom. If $m_i = m$, then the degrees of freedom is $2^k(m - 1)$.

A factorial effect in a 2^k design can be represented as

$$\bar{y}_+ - \bar{y}_-,$$

where $\bar{y}_+, \bar{y}_-$ are the averages of y for the $+, -$ levels of the factorial effect of interest (see Section 6.2.4).

Suppose that $\bar{y}_+ - \bar{y}_-$ corresponds to the BC interaction in a 2^3 design (see Table 6.3):

$$\bar{y}_+ = \frac{1}{4}\left(\bar{y}_1 + \bar{y}_2 + \bar{y}_7 + \bar{y}_8\right)$$

$$\bar{y}_- = \frac{1}{4}\left(\bar{y}_3 + \bar{y}_4 + \bar{y}_5 + \bar{y}_6\right).$$

The variance of the BC interaction from Table 6.3 is

$$
\begin{aligned}
Var(\bar{y}_+ - \bar{y}_-) &= Var(\bar{y}_+ - \bar{y}_-) \\
&= Var(\bar{y}_+) + Var(\bar{y}_-) \\
&= \frac{1}{4^2}(4\sigma^2/2 + 4\sigma^2/2) \\
&= \frac{\sigma^2}{4}.
\end{aligned}
\tag{6.2}
$$

Equation (6.2) is the variance for a factorial effect in a 2^3 design replicated twice. All factorial effects have the same standard error.

This can be estimated by substituting S^2 for σ^2 where S^2 is the pooled estimate of the 2^3 variances in (6.1); the degrees of freedom for S^2 is $2^3(2 - 1) = 8$. The standard error, the square root of the variance estimator, $s.e.(\bar{y}_+ - \bar{y}_-) = \sqrt{(1/4)S^2} = (1/2)S$.

If we assume that y_{ij} are i.i.d. and follow a normal distribution with variance σ^2, then under the null hypothesis that the factorial effect is 0

$$(\bar{y}_+ - \bar{y}_-)/(S/2) \sim t_8.$$

The variance of a factorial effect, $(\bar{y}_+ - \bar{y}_-)$ in a 2^k design replicated m times is

$$\frac{4\sigma^2}{m2^k}.\tag{6.3}$$

Under the assumption that y_{ij} are i.i.d. and follow a normal distribution with variance σ^2 then under the null hypothesis that the factorial effect is 0

$$t_{\text{eff}} = (\bar{y}_+ - \bar{y}_-) \Big/ \sqrt{\frac{4s^2}{m2^k}} \sim t_{2^k(m-1)}.\tag{6.4}$$

The distribution of t_{eff} (6.4) forms the basis for tests and confidence intervals of factorial effects. The p-value of a two-sided test that a factorial effect is zero is

$$P\left(\left|t_{2^k(m-1)}\right| > \left|t_{\text{eff}}^{obs}\right|\right),$$

where t_{eff}^{obs} is the observed value of t_{eff}.

A $100(1-\alpha)\%$ confidence interval is,

$$\bar{y}_+ - \bar{y}_- \pm \sqrt{\frac{4s^2}{m2^k}}\, t_{2^k(m-1),1-\alpha/2},$$

where $t_{2^k(m-1),1-\alpha/2}$ is the $1-\alpha/2$ quantile of the $t_{2^k(m-1)}$ distribution.

TABLE 6.3: Model Matrix for 2^3 and Data for Two Replications

Run	A	B	C	AB	AC	BC	ABC	Rep1	Rep2	Var
				Model Matrix				Data		
1	-1	-1	-1	1	1	1	-1	y_{11}	y_{12}	s_1^2
2	1	-1	-1	-1	-1	1	1	y_{21}	y_{22}	s_2^2
3	-1	1	-1	-1	1	-1	1	y_{31}	y_{32}	s_3^2
4	1	1	-1	1	-1	-1	-1	y_{41}	y_{42}	s_4^2
5	-1	-1	1	1	-1	-1	1	y_{51}	y_{52}	s_5^2
6	1	-1	1	-1	1	-1	-1	y_{61}	y_{62}	s_6^2
7	-1	1	1	-1	-1	1	-1	y_{71}	y_{72}	s_7^2
8	1	1	1	1	1	1	1	y_{81}	y_{82}	s_8^2

Example 6.2 (Silk Study). Silk fibroin is a protein that has many biomedical applications. Bucciarelli et al. [2021] used a 2^4 factorial design to study extraction of silk fibroin from silk filament. The design was replicated three times. The four factors considered are described in Table 6.4. The number of baths is a discrete variable. Time, temperature, and salt concentration are all continuous variables. The $+1$ and -1 levels correspond to the two levels of each factor used in the study. Weight loss and amount of sericin were used to determine the effectiveness of the process. The 2^4 different extractions (experimental runs) are listed in Table 6.5. Each extraction required degumming in a boiling bath for 30 minutes, rinsing three times with distilled water at room temperature then drying for two days. The order of the preparation was randomized to mediate the effect of humidity and temperature.

TABLE 6.4: Factors Tested in Bucciarelli et al. [2021]

Factor	Variable	+1 level	-1 level
A	Number of baths	1	2
B	Time	20	90
C	Temperature (celsius)	98	70
D	Salt Concentration (g/mL)	0.1	1.1

TABLE 6.5: Experimental Conditions for Example 6.2

A:Number of baths	B:Time	C:Temperature	D:Concentration
Level 1 of A	20	70	0.1
Level 2 of A	20	70	0.1
Level 1 of A	90	70	0.1
Level 2 of A	90	70	0.1
Level 1 of A	20	98	0.1
Level 2 of A	20	98	0.1
Level 1 of A	90	98	0.1
Level 2 of A	90	98	0.1
Level 1 of A	20	70	1.1
Level 2 of A	20	70	1.1
Level 1 of A	90	70	1.1
Level 2 of A	90	70	1.1
Level 1 of A	20	98	1.1
Level 2 of A	20	98	1.1
Level 1 of A	90	98	1.1
Level 2 of A	90	98	1.1

6.3.1   Computation Lab: Replication in Factorial Designs

The `silkdat` data frame contains the data from Example 6.2. `Std` is the standard order of the runs, `Run` is the randomized order in which the experimental trials were performed. The next four columns are the four factors (see Table 6.4) and the next four columns correspond to the outcomes weight loss (`mass_change`, `mass_change_pct`), and amount of sericin removed (`removed_sericin_single`, `removed_sericin_double`, `removed_sericin_pct`). For example, a weight loss of at least 26% was considered effective.

```
glimpse(silkdat)
```

```
## Rows: 48
## Columns: 11
## $ Std                    <dbl> 1, 2, 3, 4, 5, 6, 7, 8~
## $ Run                    <dbl> 47, 28, 14, 6, 33, 48,~
## $ `A:Number of baths`    <chr> "Level 1 of A", "Level~
## $ `B:Time`               <dbl> 20, 20, 20, 20, 20, 20~
## $ `C:Temperature`        <dbl> 70, 70, 70, 70, 70, 70~
## $ `D:Concentration`      <dbl> 0.1, 0.1, 0.1, 0.1, 0.~
## $ mass_change            <dbl> 0.204, 0.106, 0.076, 0~
## $ mass_change_pct        <dbl> 6.800, 3.545, 2.542, 2~
## $ removed_sericin_single <dbl> 107.2, 121.0, 140.2, 1~
## $ removed_sericin_double <dbl> 3.574, 4.046, 4.689, 4~
## $ removed_sericin_pct    <dbl> 21.200, 24.455, 25.458~
```

We can compute the model matrix for this design by recoding the four factors into variables that have values +1 and -1 for high and low values.

```
silkdat <-
  silkdat %>%
  mutate(
    A = recode(
      `A:Number of baths`,
      "Level 1 of A" = 1,
      "Level 2 of A" = -1
    ),
    B = recode(`B:Time`, `20` = 1, `90` = -1),
    C = recode(`C:Temperature`, `70` = -1, `98` = 1),
    D = recode(`D:Concentration`, `0.1` = 1, `1.1` = -1)
  )
```

There are 2^4 experimental conditions replicated three times, and it will be convenient to have a variable in `silkdat` that records each unique condition. We can select the `distinct()` experimental conditions A, B, C, and D, create a new variable `exp_cond` from the `row_number()`, and then `right_join()` this data frame with the original data frame.

```
silkdat <-
  silkdat %>%
  select(A, B, C, D) %>%
  distinct() %>%
  mutate(exp_cond = row_number()) %>%
  right_join(silkdat, by = c("A", "B", "C", "D"))
```

We will focus on the percentage of mass loss after the process `mass_change_pct`. Bucciarelli et al. [2021] considered runs that had a percent mass change at least 26% a success. We can examine the mean of `mass_change_pct` for the three replications for each experimental condition. This allows us to quickly examine the effects of different experimental conditions on percentage of mass loss. For example, experimental conditions 11 and 12 are identical except factor A is at its low level in condition 11 and high level in condition 12, and condition 12 leads to a process that is effective, while condition 11 does not.

```
silkdat %>%
  group_by(exp_cond) %>%
  mutate(m = mean(mass_change_pct)) %>%
  ungroup() %>%
  distinct(exp_cond, A, B, C, D, m)
```

```
## # A tibble: 16 x 6
##          A     B     C     D exp_cond     m
##      <dbl> <dbl> <dbl> <dbl>    <int> <dbl>
## 1      1     1    -1     1         1  4.30
## 2     -1     1    -1     1         2  4.38
## 3      1    -1    -1     1         3  2.12
## 4     -1    -1    -1     1         4  5.64
## 5      1     1     1     1         5 16.2
## 6     -1     1     1     1         6 19.2
## 7      1    -1     1     1         7 21.8
## 8     -1    -1     1     1         8 25.8
## 9      1     1    -1    -1         9 15.1
## 10    -1     1    -1    -1        10 20.8
```

```
## 11     1    -1    -1    -1        11 22.8
## 12    -1    -1    -1    -1        12 26.7
## 13     1     1     1    -1        13 25.7
## 14    -1     1     1    -1        14 26.6
## 15     1    -1     1    -1        15 26.5
## 16    -1    -1     1    -1        16 26.8
```

We can use the function `calcmaineff()` developed in Section 6.2.5 to calculate main effects. `map_dbl` is used to iterate over a list of main effects and return a vector of main effects. Factor D has the largest effect on mass loss.

```
silk_ME <-
  list("A", "B", "C", "D") %>%
  map_dbl(function(x)
    calcmaineff(df = silkdat, y1 = "mass_change_pct", fct1 = x))
silk_ME
```

```
## [1]  -2.688  -3.243  10.844 -11.459
```

It's also possible to calculate two-way, three-way, and four-way interactions.

The variance (`s2`) and the standard error (`std_err`) of factorial effects can be calculated using (6.1).

```
silk_var <-
  silkdat %>%
  group_by(exp_cond) %>%
  mutate(v = var(mass_change_pct)) %>%
  ungroup() %>%
  distinct(exp_cond, v) %>%
  summarise(s2 = sum(v) / (2 ^ 4),
            std_err = sqrt((4 * s2) / (3 * 2 ^ 4)))

silk_var
```

```
## # A tibble: 1 x 2
##        s2 std_err
##     <dbl>   <dbl>
## 1  6.76    0.751
```

The standard errors can be used to directly calculate p-values and confidence intervals for factorial effects. For example, the calculations below show the p-values for testing that the main effects are 0 and 99% confidence intervals. The degrees of freedom is $2^4 \times (3 - 1) = 32$. Finally, the results are stored in a data frame for ease of comparison and communication.

```
t_silk <- silk_ME / silk_var$std_err

df_silk <- 2 ^ 4 * (3 - 1)

pvals <- 2 * (1 - pt(abs(t_silk), df = df_silk))

alpha <- 0.01

MError <- silk_var$std_err * qt(p = 1 - (alpha / 2), df = df_silk)

CI_L <- silk_ME - MError
CI_U <- silk_ME + MError

data.frame(ME = LETTERS[1:4],
           silk_ME,
           pvalue = round(pvals, 5),
           CI_L,
           CI_U)
```

```
##   ME silk_ME  pvalue     CI_L     CI_U
## 1  A  -2.688 0.00112  -4.743  -0.6325
## 2  B  -3.243 0.00014  -5.299  -1.1880
## 3  C  10.844 0.00000   8.788  12.8990
## 4  D -11.459 0.00000 -13.515  -9.4038
```

None of the 99% confidence intervals for the main effects include 0, which indicates that all four factors had a significant effect on percentage of weight loss.

6.4 Linear Model for a 2^k Factorial Design

The factorial effects in a 2^k design can be estimated by fitting a linear regression model. Consider the 2^3 design discussed in Table 6.2. Define

$$x_{i1} = \begin{cases} 1 & \text{if } A = 1 \\ -1 & \text{if } A = -1 \end{cases}, \quad x_{i2} = \begin{cases} 1 & \text{if } B = 1 \\ -1 & \text{if } B = -1 \end{cases}, \quad x_{i3} = \begin{cases} 1 & \text{if } C = 1 \\ -1 & \text{if } C = -1 \end{cases}$$

where $i = 1, \ldots, 8$. Let y_i be the data from the ith run. Denote the columns A, B, C by x_{i1}, x_{i2}, and x_{i3}, where $i = 1, \ldots, 8$. Then the columns corresponding to the interactions, AB, AC, BC, and ABC, can be represented by $x_{i4} = x_{i1}x_{i2}$, $x_{i5} = x_{i1}x_{i3}$, $x_{i6} = x_{i2}x_{i3}$, and $x_{i7} = x_{i1}x_{i2}x_{i3}$.

A linear model for a 2^3 factorial design is

$$y_i = \beta_0 + \sum_{j=1}^{7} \beta_i x_{ij} + \epsilon_i,$$

$i = 1, \ldots, 8$. The least squares estimates of β_j is

$$\hat{\beta}_j = \frac{1}{2}\left(\bar{y}(x_{ij} = +1) - \bar{y}(x_{ij} = -1)\right),$$

where $\bar{y}(x_{ij} = 1)$ is the mean of y_i when $x_{ij} = 1$. For example, the main effect of A is $\bar{y}(x_{i1} = +1) - y(x_{i1} = -1) = (1/4)(\sum_{\{i:x_{i1}=+1\}} y_i - \sum_{\{i:x_{i1}=-1\}} y_i) = 2\hat{\beta}_1$. Therefore, each factorial effect can be obtained by doubling the corresponding $\hat{\beta}_j$.

6.4.1   Computation Lab: Linear Model for a 2^k Factorial Design

A linear regression model can be fit to the 2^3 design in Table 6.2. The data from Example 6.1 is in the data frame wtlossdat.

```
fooddiary_mod <- lm(y ~ A * B * C, data = wtlossdat)
model.matrix(fooddiary_mod)
```

```
##    (Intercept)  A  B  C A:B A:C B:C A:B:C
## 1            1 -1 -1 -1   1   1   1    -1
## 2            1 -1 -1  1   1  -1  -1     1
## 3            1 -1  1 -1  -1   1  -1     1
## 4            1 -1  1  1  -1  -1   1    -1
## 5            1  1 -1 -1  -1  -1   1     1
## 6            1  1 -1  1  -1   1  -1    -1
## 7            1  1  1 -1   1  -1  -1    -1
## 8            1  1  1  1   1   1   1     1
## attr(,"assign")
## [1] 0 1 2 3 4 5 6 7
```

The model matrix of `fooddiary_mod` shows the **X** matrix used to calculate the $\hat{\beta} = \left(X'X\right)^{-1} X'\mathbf{y}$ (see (2.5)).

```
summary(fooddiary_mod)
```

```
##
## Call:
## lm.default(formula = y ~ A * B * C, data = wtlossdat)
##
## Residuals:
## ALL 8 residuals are 0: no residual degrees of freedom!
##
## Coefficients:
##               Estimate Std. Error t value Pr(>|t|)
## (Intercept)    0.6375         NaN     NaN      NaN
## A              0.2625         NaN     NaN      NaN
## B             -0.9875         NaN     NaN      NaN
## C              0.2875         NaN     NaN      NaN
## A:B            0.0875         NaN     NaN      NaN
## A:C            0.3125         NaN     NaN      NaN
## B:C           -0.2875         NaN     NaN      NaN
## A:B:C          0.0875         NaN     NaN      NaN
##
## Residual standard error: NaN on 0 degrees of freedom
## Multiple R-squared:     1,    Adjusted R-squared:    NaN
## F-statistic:  NaN on 7 and 0 DF,  p-value: NA
```

The standard errors of the estimates are shown as NA since the design is unreplicated; at least two observations per run are needed to estimate the standard error. Let's verify that if the least squares estimates are doubled, then they are equal to the factorial effects. For example, estimates of the main effects are:

```
map_dbl(list("A", "B", "C"),
        function(x)
            calcmaineff(df = wtlossdat, y1 = "y", fct1 = x))
```

```
## [1]   0.525 -1.975   0.575
```

Twice the least squares estimates corresponding to the main effects is

```
2 * coef(fooddiary_mod)[2:4]
```

```
##     A      B      C
## 0.525 -1.975  0.575
```

Linear regression can be used to estimate p-values and confidence intervals for factorial effects in *replicated designs*. A linear regression model of the replicated 2^4 design in Example 6.2 is

```
silk_mod <- lm(mass_change_pct ~ A * B * C * D, data = silkdat)
```

A summary of the model shows the p-values for the factorial effects.

```
summary(silk_mod)
```

```
##
## Call:
## lm.default(formula = mass_change_pct ~ A * B * C * D,
##     data = silkdat)
##
## Residuals:
##     Min     1Q Median     3Q    Max
## -4.189 -1.237 -0.083  0.838  6.578
##
## Coefficients:
##               Estimate Std. Error t value Pr(>|t|)
## (Intercept)    18.1571     0.3753   48.38  < 2e-16 ***
## A              -1.3440     0.3753   -3.58  0.00112 **
## B              -1.6217     0.3753   -4.32  0.00014 ***
## C               5.4218     0.3753   14.45  1.4e-15 ***
## D              -5.7296     0.3753  -15.27  3.0e-16 ***
## A:B             0.1242     0.3753    0.33  0.74291
## A:C             0.3113     0.3753    0.83  0.41304
## B:C            -0.0287     0.3753   -0.08  0.93958
## A:D             0.0128     0.3753    0.03  0.97296
## B:D             0.2044     0.3753    0.54  0.58976
## C:D             2.8944     0.3753    7.71  8.6e-09 ***
## A:B:C          -0.0643     0.3753   -0.17  0.86496
```

```
## A:B:D          0.4360        0.3753     1.16  0.25390
## A:C:D         -0.7408        0.3753    -1.97  0.05706 .
## B:C:D         -1.6159        0.3753    -4.31  0.00015 ***
## A:B:C:D       -0.2341        0.3753    -0.62  0.53725
## ---
## Signif. codes:
## 0 '***' 0.001 '**' 0.01 '*' 0.05 '.' 0.1 ' ' 1
##
## Residual standard error: 2.6 on 32 degrees of freedom
## Multiple R-squared:  0.946,  Adjusted R-squared:  0.92
## F-statistic: 37.2 on 15 and 32 DF,  p-value: 5.45e-16
```

The factorial effects can be obtained by doubling the estimated regression coefficients, and $100(1-\alpha)\%$ confidence intervals are computed by doubling the lower and upper limits of the confidence intervals for the regression coefficients.

```
2 * coef(silk_mod)
```

```
## (Intercept)            A            B            C
##    36.31429     -2.68797     -3.24339     10.84358
##           D          A:B          A:C          B:C
##   -11.45927      0.24833      0.62250     -0.05734
##         A:D          B:D          C:D        A:B:C
##     0.02564      0.40881      5.78872     -0.12867
##       A:B:D        A:C:D        B:C:D      A:B:C:D
##     0.87203     -1.48164     -3.23184     -0.46812
```

```
2 * confint(silk_mod, level = 0.99)
```

```
##                 0.5 %   99.5 %
## (Intercept)    34.259  38.3697
## A             -4.743  -0.6325
## B             -5.299  -1.1880
## C              8.788  12.8990
## D            -13.515  -9.4038
## A:B           -1.807   2.3038
## A:C           -1.433   2.6779
## B:C           -2.113   1.9981
## A:D           -2.030   2.0811
## B:D           -1.647   2.4642
## C:D            3.733   7.8441
```

```
## A:B:C        -2.184   1.9268
## A:B:D        -1.183   2.9275
## A:C:D        -3.537   0.5738
## B:C:D        -5.287  -1.1764
## A:B:C:D       -2.524   1.5873
```

6.5   Normal Plots in Unreplicated Factorial Designs

Normal quantile plots (see Section 2.5) can be used to assess the normality of a set of data.

Let $X_{(1)} < ... < X_{(N)}$ denote the ordered values of random variables $X_1, ..., X_N$. For example, $X_{(1)}$ and $X_{(N)}$ are the minimum and maximum of $X_1, ..., X_N$.

Normal quantile plots are constructed based on the following logic:

1. $\Phi(X_{(i)})$ has a uniform distribution on $[0, 1]$.

2. $E(\Phi(X_{(i)})) = i/(N + 1)$ (the expected value of the ith order statistic from a Unif$[0, 1]$).

3. 2 implies that the N points $(p_i, \Phi(X_{(i)}))$, where $p_i = i/(N + 1)$, should fall on a straight line. Applying the Φ^{-1} transformation to the horizontal and vertical scales, the N points
$$\left(\Phi^{-1}(p_i), X_{(i)}\right) \tag{6.5}$$

   form the normal quantile plot of $X_1, ..., X_N$. If $X_1, ..., X_N$ are generated from a normal distribution, then a plot of the points $\left(\Phi^{-1}(p_i), X_{(i)}\right)$, $i = 1, ..., N$ should be a straight line.

4. The points (6.5) are compared to the straight line that passes through the first and third quartiles of the ordered sample and standard normal distribution.

A marked (systematic) deviation of the plot from the straight line would indicate that either:

1. The normality assumption does not hold.
2. The variance is not constant.

These plots can be used to evaluate the normality of the effects in a 2^k factorial design. Let $\hat{\theta}_{(1)} < \hat{\theta}_{(2)} < \cdots < \hat{\theta}_{(N)}$ be N ordered factorial estimates. If we plot the points

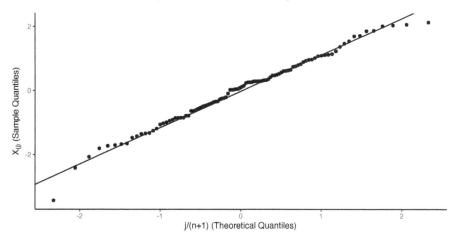

FIGURE 6.6: Normal Quantile Plot of 100 Simulated Observations from $N(0, 1)$

$$\left(\hat{\theta}_i,\ \Phi^{-1}(p_i)\right),\ i = 1, ..., N,$$

where, $p_i = i/(N + 1)$, then factorial effects $\hat{\theta}_i$ that are close to 0 will fall along a straight line. Therefore, points that fall off the straight line will be declared significant.

The rationale is as follows:

1. Assume that the estimated effects $\hat{\theta}_i$ are $N(\theta, \sigma^2)$ (estimated effects involve averaging of N observations, and the CLT (Section 2.6) ensures averages are nearly normal, even for N as small as 8).

2. If $H_0 : \theta_i = 0$, $i = 1, ..., N$ is true, then all the estimated effects will be zero.

3. The resulting normal probability plot of the estimated effects will be a straight line.

4. Therefore, the normal quantile plot is testing whether all of the estimated effects have the same distribution (i.e. same means).

When some of the effects are nonzero, the corresponding estimated effects will tend to be larger and fall off the straight line. For positive effects, the estimates fall above the line and negative effects fall below the line. Figure 6.7 shows a normal quantile plot of the factorial effects for Example 6.1, where the points are labeled as the main effects, A, B, C, and interactions A:B, B:C, A:C, A:B:C.

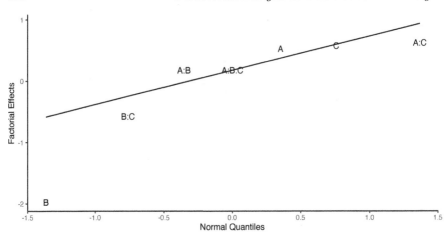

FIGURE 6.7: Normal Plot for Example 6.1

6.5.1   Half-Normal Plots

A related graphical method is called the half-normal probability plot. Let

$$\left|\hat{\theta}\right|_{(1)} < \left|\hat{\theta}\right|_{(2)} < \cdots < \left|\hat{\theta}\right|_{(N)}$$

be the ordered values of the absolute value of the factorial effect estimates. A normal quantile plot of the absolute value of factorial effects or half-normal plot can also be constructed.

Assume that the estimated effects $\hat{\theta}_i$ are $N(0, \sigma^2)$, so $|\hat{\theta}_i|$ has p.d.f.

$$\frac{\sqrt{2}}{\sigma\sqrt{\pi}} \exp\left(-y^2/2\sigma^2\right), \, y \geq 0.$$

This is the p.d.f. of the half-normal distribution.

Suppose Y has a half-normal distribution with $\sigma^2 = 1$. Let y_p be the pth quantile of Y, so $p = F(y_p) = P(Y \leq y_p)$. It can be shown that $F(y_p) = 2(\Phi(y_p) - 1/2))$. This implies that

$$y_p = F^{-1}(p) = \Phi^{-1}(p/2 + 1/2).$$

A quantile - quantile plot of the absolute values of factorial effects can use the standard normal CDF to compute the theoretical quantiles of the half-normal distribution.

The half-normal probability plot consists of the points

$$\left(\left|\hat{\theta}\right|_{(i)}, \Phi^{-1}(0.5 + 0.5[i - 0.5]/N)\right), \, i = 1, ..., N.$$

This plot uses the first and third quantiles of the half-normal distribution to construct a comparative straight line.

An advantage of this plot is that all the large estimated effects appear in the upper right-hand corner and fall above the line. Figure 6.8 shows a half-normal plot of the factorial effects for Example 6.1. Normal plots can often be misleading. For example, Figure 6.7 suggests that the BC interaction might be interesting, even though it is smaller in magnitude than the AB effect. The half-normal plot avoids this visual trap. In this case, it shows that B is the only important effect.

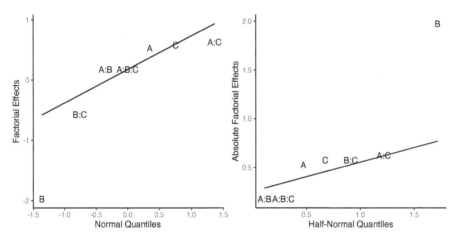

FIGURE 6.8: Full Normal and Half-Normal Plot for Example 6.1

6.5.2 Computation Lab: Normal Plots in Unreplicated Factorial Designs

Consider the data from the 2^3 design in Example 6.1. The factorial effects were estimated in Example 6.1 via a linear regression model in Section 6.4.1 and stored in `fooddiary_mod`. A vector of ordered factorial effects can be obtained using `dplyr::arrange()`.

The normal quantile plot of factorial effects in Figure 6.7 was created using `ggplot`. To create the plot using the factor labels, we pass `geom = "text"` and `label = rownames(effs)` to `ggplot2::geom_qq()`. This tells `ggplot2::geom_qq()` to draw text using the `rownames()` of `effs` instead of the default `geom = point`. Finally, `distribution = stats::qnorm` specifies the distribution function to use.

```
ggplot(effs, aes(sample = theta)) +
  geom_qq(
    geom = "text",
    label = rownames(effs),
    distribution = stats::qnorm
  ) +
  geom_qq_line(distribution = stats::qnorm) +
  ylab("Factorial Effects") +
  xlab("Normal Quantiles")
```

A half-normal plot of the factorial effects can be created using ggplot() by computing the ordered absolute values of the factorial effects, and the corresponding theoretical quantiles of the half-normal and storing these values in a data frame.

```
effs <- data.frame(theta = abs(2 * coef(fooddiary_mod)[-1]))

effs <-
  effs %>%
  arrange(theta) %>%
  mutate(pi = qnorm(0.5 +
                    0.5 * (1:length(effs$theta) - 0.5) /
                    length(effs$theta)))
```

Instead of using ggplot2::geom_qq() and ggplot2::geom_qq_line(), geom_text() is used with label = rownames(effs) to draw the names of the factors.

```
ggplot(effs, aes(pi, theta)) +
  geom_text(label = rownames(effs)) +
  ylab("Absolute Factorial Effects") +
  xlab("half-Normal Quantiles")
```

Finally, a straight line is constructed that connects the first and third quartiles of the sample and theoretical quartiles. The quartile function of the half-normal distribution qhalf() is defined in terms of qnorm() and used to compute the intercept and slope of the straight line.

```
qhalf <- function(p) {
  qnorm(0.5 + 0.5 * p)
}

q <- quantile(effs$theta, c(0.25, 0.75))
slope <- (q[2] - q[1]) / (qhalf(.75) - qhalf(.25))
int <- q[1] - slope * qhalf(.25)
```

A reference line can be added to the plot using `ggplot2::geom_abline()`.

```
ggplot(effs, aes(pi, theta)) +
  geom_text(label = rownames(effs)) +
  ylab("Absolute Factorial Effects") + xlab("half-Normal Quantiles") +
  geom_abline(intercept = int, slope = slope)
```

This code produces Figure 6.8.

`BsMD::DanielPlot()` creates normal and half-normal plots, although there is no option to add a reference line at this time.

6.6   Lenth's Method

Half-normal and normal plots are informal graphical methods involving visual judgment. It's desirable to judge a deviation from a straight line quantitatively based on a formal test of significance. Lenth [1989] proposed a method that is simple to compute and performs well.

Let

$$\hat{\theta}_1, ..., \hat{\theta}_N$$

be estimated factorial effects of $\theta_1, \theta_2, ..., \theta_N$. Assume that all the factorial effects have the same standard deviation.

The pseudo standard error (PSE) is defined as

$$PSE = 1.5 \cdot \text{median}_{|\hat{\theta}_i| < 2.5s_0} \left| \hat{\theta}_i \right|,$$

where the median is computed among the $\left|\hat{\theta}_i\right|$ with $\left|\hat{\theta}_i\right| < 2.5 s_0$ and

$$s_0 = 1.5 \cdot \text{median} \left|\hat{\theta}_i\right|.$$

$1.5 \cdot s_0$ is a consistent estimator of the standard deviation of $\hat{\theta}$ when $\theta_i = 0$ and the underlying distribution is normal. The $P(|Z| > 2.57) = 0.01$, $Z \sim N(0,1)$. So, $\left|\hat{\theta}_i\right| < 2.5 s_0$ trims approximately 1% of the $\hat{\theta}_i$ if $\theta_i = 0$. The trimming attempts to remove the $\hat{\theta}_i$ associated with non-zero (active) effects. Using the median in combination with the trimming means that PSE is not sensitive to the $\hat{\theta}_i$ associated with active effects.

To obtain a margin of error Lenth suggested multiplying the PSE by the $100 * (1 - \alpha)$ quantile of the t_d distribution, $t_{d,\alpha/2}$. The degrees of freedom is $d = m/3$, where m is the number of factorial estimates [Lenth, 1989]. For example, the margin of error for a 95% confidence interval for θ_i is

$$ME = t_{d,.025} \times PSE.$$

All estimates greater than the ME may be viewed as "significant," but with so many estimates being considered simultaneously, some may be falsely identified.

A simultaneous margin of error that accounts for multiple testing can also be calculated,

$$SME = t_{d,\gamma} \times PSE,$$

where $\gamma = \left(1 + 0.95^{1/N}\right)/2$.

6.6.1 Computation Lab: Lenth's method

In this section, we will use Lenth's method to test effect significance in the unreplicated 2^3 design presented in Example 6.1.

Let's calculate Lenth's method for the process development example. The estimated factorial effects are:

```
effs <- data.frame(theta = 2 * coef(fooddiary_mod)[-1])
round(effs, 2)
```

```
##          theta
## A         0.53
## B        -1.97
## C         0.58
```

```
## A:B      0.17
## A:C      0.62
## B:C     -0.58
## A:B:C   0.18
```

The estimate of $s_0 = 1.5 \cdot \text{median} \left| \hat{\theta}_i \right|$ is

```
s0 <- 1.5 * median(abs(effs$theta))
s0
```

```
## [1] 0.8625
```

The trimming constant $2.5s_0$ is

```
2.5 * s0
```

```
## [1] 2.156
```

The effects $\hat{\theta}_i$ such that $\left| \hat{\theta}_i \right| \geq 2.5s_0$ will be trimmed. Below we mark the non-trimmed effects TRUE

```
cond <- abs(effs$theta) < 2.5 * s0
cond
```

```
## [1] TRUE TRUE TRUE TRUE TRUE TRUE TRUE
```

The *PSE* is then calculated as 1.5 times the median of these values.

```
PSE <- 1.5 * median(abs(effs$theta[cond]))
PSE
```

```
## [1] 0.8625
```

The *ME* and SME are

```
df_foodmod <- (2^3 - 1) / 3
```

```
ME <- PSE * qt(p = .975, df = df_foodmod)
ME
```

```
## [1] 3.247
```

```
SME <- PSE * qt(p = (1 + .95 ^ {
  1 / (8 - 1)
}) / 2, df = df_foodmod)
SME
```

```
## [1] 7.77
```

So, 95% confidence intervals, using the individual error rate, for the effects are

```
CI_L <- round(effs$theta - ME, 2)
CI_U <- round(effs$theta + ME, 2)
cbind(factor = rownames(effs),
      effects = round(effs$theta, 2),
      CI_L,
      CI_U)
```

```
##        factor  effects CI_L    CI_U
## [1,]  "A"      "0.53"  "-2.72" "3.77"
## [2,]  "B"      "-1.97" "-5.22" "1.27"
## [3,]  "C"      "0.58"  "-2.67" "3.82"
## [4,]  "A:B"    "0.17"  "-3.07" "3.42"
## [5,]  "A:C"    "0.62"  "-2.62" "3.87"
## [6,]  "B:C"    "-0.58" "-3.82" "2.67"
## [7,]  "A:B:C"  "0.18"  "-3.07" "3.42"
```

A plot of the effects with a ME and SME is usually called a Lenth plot. In R, it can be implemented via the function **Lenthplot()** in the **BsMD** library. The values of PSE, ME, and SME are part of the output. The spikes in the plot below are used to display factor effects.

BsMD::LenthPlot(fooddiary_mod)

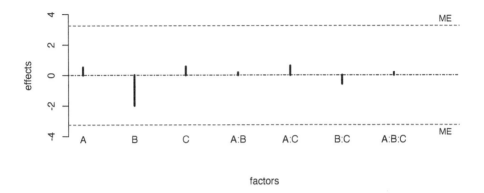

```
## alpha     PSE     ME     SME
## 0.0500 0.8625 3.2466 7.7697
```

6.7 Blocking Factorial Designs

Blocking is usually used to balance the design with respect to a factor that may influence the outcome of interest and doesn't interact with the experimental factors of interest. In the randomized block design discussed in Section 3.10, all treatments are compared in each block. This is feasible when the number of treatments doesn't exceed the block size. In a 2^k design, this is usually not realistic. One approach is to assign a fraction of the runs in a 2^k design to each block.

Consider a 2^3 design that uses batches of raw material to create the treatment combinations, but the batches of raw material were only large enough to complete four runs of the experiment. Table 6.6 shows the design matrix for a 2^3 design in standard order, where the factors are labelled **1**, **2**, **3**, and the interaction between factors **1** and **2** labelled as **12**, etc. The levels of the variables are shown as − and + instead of +1 and −1.

TABLE 6.6: Arranging a 2^3 Design in Two Blocks of Size 4 Using Three-Way Interaction

Run	1	2	3	12	13	23	4=123	Block
1	−	−	−	+	+	+	−	1

2	+	−	−	−	−	+	+	2
3	−	+	−	−	+	−	+	2
4	+	+	−	+	−	−	−	1
5	−	−	+	+	−	−	+	2
6	+	−	+	−	+	−	−	1
7	−	+	+	−	−	+	−	1
8	+	+	+	+	+	+	+	2

Suppose that we assign runs 1, 4, 6, and 7 to block 1 which uses the first batch of raw material and runs 2, 3, 5, and 8 to block 2 which uses the second batch of raw material. The design is blocked this way by placing all runs in which the three-way interaction **123** is minus in one block and all the other runs in which **123** is plus in the other block (see Table 6.6).

Any systematic differences between the two blocks of four runs will be eliminated from all the main effects and two factor interactions. What you gain is the elimination of systematic differences between blocks. But now the three factor interaction is confounded with any batch (block) difference. The ability to estimate the three factor interaction separately from the block effect is lost.

6.7.1   Effect Heirarchy Principle

1. Lower-order effects are more likely to be important than higher-order effects.

2. Effects of the same order are equally likely to be important.

This principle suggests that when resources are scarce, priority should be given to the estimation of lower order effects. This is useful in screening experiments that have a large number of factors and a relatively small number of runs. Section 6.8 will discuss factorial designs that can study a large number of factors in a fraction of the runs needed in a full factorial design.

One reason that many accept this principle is that higher order interactions are more difficult to interpret or justify physically. As a result, investigators are less interested in estimating the magnitudes of these effects even when they are statistically significant.

Assigning a fraction of the 2^k treatment combinations to each block results in an incomplete blocking (e.g., balanced incomplete block design). The factorial structure of a 2^k design allows a neater assignment of treatment combinations to blocks. The neater assignment is done by dividing the total combinations into various fractions and finding optimal assignments by exploiting combinatorial relationships.

6.7.2 Generation of Orthogonal Blocks

In the 2^3 example in Table 6.6, the blocking variable is given the identifying number **4**. Think of the study as containing four factors, where the fourth factor will have the special property that it does not interact with other factors. If this new factor is introduced by having its levels coincide exactly with the plus and minus signs attributed to **123**, then the blocking is said to be *generated by the relationship* **4 = 123**. This idea can be used to derive more sophisticated blocking arrangements.

Example 6.3. The following example is from Box et al. [2005]. Consider an arrangement of a 2^3 design into four blocks using two block factors called 4 and 5 (see Table 6.8). 4 is associated with the three factor interaction, and 5 is associated with a the two factor interaction 23 which was deemed unimportant by the investigator. Runs are placed in different blocks depending on the signs of the block variables in columns 4 and 5. Runs for which the signs of 4 and 5 are −− would go in one block, −+ in a second block, +− in a third block, and ++ in the fourth.

Block	Run
I	4,6
II	3,5
III	1,7
IV	2,8

Block variables 4 and 5 are confounded with interactions 123 and 23. But there are three degrees of freedom associated with four blocks. The third degree of freedom accommodates the 45 interaction. But, the 45 interaction has the same signs as the main effect 1, since 45 = 1. Therefore, if we use 4 and 5 as blocking variables, then it will be confounded with block differences (see Table 6.8).

Main effects should not be confounded with block effects, and any blocking scheme that confounds main effects with blocks should not be used. This is based on the assumption that block-by-treatment interactions are negligible.

This assumption states that treatment effects do not vary from block to block. Without this assumption, estimability of the factorial effects will be very complicated.

For example, if $B_1 = 12$ then this implies two other relations:

$$1B_1 = 1 \times B_1 = 112 = 2 \text{ and } B_1 2 = B_1 \times 2 = 122 = 1.$$

If there is a significant interaction between the block effect B_1 and the main effect 1, then the main effect 2, is confounded with $B_1 1$. Similarly, if there is a significant interaction between the block effect B_1 and the main effect 2 then the main effect 1 is confounded with $B_1 2$. Interactions can be checked by plotting the residuals for all the treatments within each block. If the pattern varies from block to block, then the assumption may be violated. A block-by-treatment interaction often suggests interesting information about the treatment and blocking variables.

TABLE 6.8: 2^3 Arranged in Four Blocks

run	1	2	3	12	13	5=23	4=123	45=1
1	−	−	−	+	+	+	−	−
2	+	−	−	−	−	+	+	+
3	−	+	−	−	+	−	+	−
4	+	+	−	+	−	−	−	+
5	−	−	+	+	−	−	+	−
6	+	−	+	−	+	−	−	+
7	−	+	+	−	−	+	−	−
8	+	+	+	+	+	+	+	+

6.7.3   Generators and Defining Relations

A simple calculus is available to show the consequences of any proposed blocking arrangement. If any column in a 2^k design is multiplied by itself, a column of plus signs is obtained. This is denoted by the symbol I. Thus, you can write

$$I = 11 = 22 = 33 = 44 = 55,$$

where, for example, 22 means the product of the elements of column 2 with itself.

Any column multiplied by I leaves the elements unchanged. So, $I3 = 3$.

A general approach for arranging a 2^k design in 2^q blocks of size 2^{k-q} is as follows. Let $B_1, B_2, ..., B_q$ be the block variables, and the factorial effect v_i is confounded with B_i,

$$B_1 = v_1, B_2 = v_2, ..., B_q = v_q.$$

The block effects are obtained by multiplying the B_i's:

$$B_1 B_2 = v_1 v_2, B_1 B_3 = v_1 v_3, ..., B_1 B_2 \cdots B_q = v_1 v_2 \cdots v_q$$

There are $2^q - 1$ possible products of the B_i's and the I (whose components are $+$).

Example 6.4. A 2^5 design can be arranged in eight blocks of size four.

Consider two blocking schemes.

Blocking scheme 1: Define the blocks as

$$B_1 = 135, B_2 = 235, B_3 = 1234.$$

The remaining blocks are confounded with the following interactions:

$$B_1B_2 = 12, B_1B_3 = 245, B_2B_3 = 145, B_1B_2B_3 = 34$$

In this blocking scheme, the seven block effects are confounded with the seven interactions

$$12, 34, 135, 145, 235, 245, 1234.$$

Blocking scheme 2:

$$B_1 = 12, B_2 = 13, B_3 = 45.$$

This blocking scheme confounds the following interactions.

$$12, 13, 23, 45, 1245, 1345, 2345.$$

Which is a better blocking scheme?

The second scheme confounds four two-factor interactions, while the first confounds only two two-factor interactions. Since two-factor interactions are more likely to be important than three- or four-factor interactions, the first scheme is superior.

6.7.4 Computation Lab: Blocking Factorial Designs

A 2^3 design in standard order can be generated by `FrF2::FrF2()`. `nruns` is the number of runs which is $2^3 = 8$ in this case; the number of factors `nfactors` is 3, and `randomize = FALSE` (this tells `FrF2()` to keep runs in standard order; if `randomize = TRUE` then the order of the runs is randomized, i.e., not in standard order).

```
FrF2::FrF2(nruns = 8,
           nfactors = 3,
           randomize = FALSE)
```

```
##     A  B  C
## 1 -1 -1 -1
## 2  1 -1 -1
## 3 -1  1 -1
## 4  1  1 -1
## 5 -1 -1  1
## 6  1 -1  1
## 7 -1  1  1
## 8  1  1  1
## class=design, type= full factorial
```

A 2^k design generated by 2^q blocks of size 2^{k-q} can be generated using FrF2()
using the **blocks** parameter. For example, a 2^5 design can be arranged in 2^3
blocks by defining blocks using the three independent variables $\mathbf{B_1 = 135}, \mathbf{B_2 =}$
235, and $\mathbf{B_3 = 1234}$. FrF2 factors are labelled with capital letters starting
from the beginning of the alphabet.

```
FrF2::FrF2(
  nruns = 32,
  nfactors = 5,
  randomize = F,
  blocks = c("ACE", "BCE", "ABCD")
)
```

This design is aliased or confounded with two two-factor interactions, and
FrF2() will give an error message in this case. Specifying **alias.block.2fis**
= TRUE will generate the design, and the aliasing information is available
through **aliased.with.blocks**.

```
d325 <- FrF2::FrF2(
  nruns = 32,
  nfactors = 5,
  randomize = F,
  blocks = c("ACE", "BCE", "ABCD"),
  alias.block.2fis = TRUE
```

```
)
design.info(d325)$aliased.with.blocks
```

```
## [1] "AB" "CD"
```

Consider an alternative blocking scheme, where $\mathbf{B_1}$ = 12, $\mathbf{B_2}$ = 13, and $\mathbf{B_3}$ = 45.

```
d325 <- FrF2::FrF2(
  nruns = 32,
  nfactors = 5,
  randomize = F,
  blocks = c("AB", "AC", "DE"),
  alias.block.2fis = TRUE
)
design.info(d325)$aliased.with.blocks
```

```
## [1] "AB" "AC" "BC" "DE"
```

This blocking scheme is confounded with four two-factor interactions. Therefore, the first blocking scheme is preferred.

6.8 Fractional Factorial Designs

A 2^k full factorial requires 2^k runs. Full factorials are seldom used in practice for large k (k>=7). For economic reasons fractional factorial designs, which consist of a fraction of full factorial designs, are used. There are criteria to choose "optimal" fractions.

Consider a design that studies six factors in 32 (2^5) runs, which is a $\frac{1}{2}$ fraction of a 2^6 factorial design. The design is referred to as a 2^{6-1} design, where 6 denotes the number of factors and 2^{-1} denotes the fraction. An important issue that arises in the design of these studies is how to select the fraction.

Example 6.5 (Treating HSV-1 with Drug Combinations). Jaynes et al. [2013] used a 2^{6-1} fractional factorial design to a investigate a biological system with HSV-1 (Herpes simplex virus type 1) and six antiviral drugs. All combinations of six drugs, A, B, C, D, E, and F, were studied in 32 runs. The low and high levels of each drug were coded as −1 and +1. Table 6.9 displays the design matrix and data. The levels of drug F is the product of A, B, C, D, and E, that is, $F = ABCDE$. The outcome variable, readout, is the percentage of cells positive for HSV-1 after combinatorial drug treatments.

TABLE 6.9: Design Matrix for 2^{6-1} with $F = ABCDE$

Run	A	B	C	D	E	F	Readout
1	-1	-1	-1	-1	-1	-1	31.6
2	-1	-1	-1	-1	1	1	32.6
3	-1	-1	-1	1	-1	1	13.4
4	-1	-1	-1	1	1	-1	13.2
5	-1	-1	1	-1	-1	1	27.5
6	-1	-1	1	-1	1	-1	32.5
7	-1	-1	1	1	-1	-1	11.6
8	-1	-1	1	1	1	1	20.8
9	-1	1	-1	-1	-1	1	37.2
10	-1	1	-1	-1	1	-1	51.6
11	-1	1	-1	1	-1	-1	14.1
12	-1	1	-1	1	1	1	19.9
13	-1	1	1	-1	-1	-1	27.3
14	-1	1	1	-1	1	1	40.2
15	-1	1	1	1	-1	1	19.3
16	-1	1	1	1	1	-1	23.3
17	1	-1	-1	-1	-1	1	31.2
18	1	-1	-1	-1	1	-1	32.6
19	1	-1	-1	1	-1	-1	14.2
20	1	-1	-1	1	1	1	22.4
21	1	-1	1	-1	-1	-1	32.7
22	1	-1	1	-1	1	1	41.0
23	1	-1	1	1	-1	1	20.1
24	1	-1	1	1	1	-1	18.7
25	1	1	-1	-1	-1	-1	29.6
26	1	1	-1	-1	1	1	42.3
27	1	1	-1	1	-1	1	18.5
28	1	1	-1	1	1	-1	20.0
29	1	1	1	-1	-1	1	30.9
30	1	1	1	-1	1	-1	34.3
31	1	1	1	1	-1	-1	19.4
32	1	1	1	1	1	1	23.4

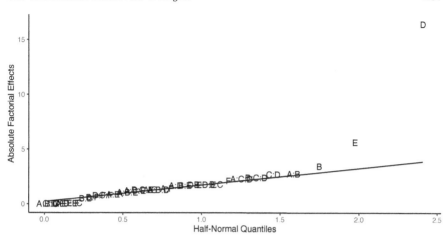

FIGURE 6.9: Half-Normal for Example 6.5

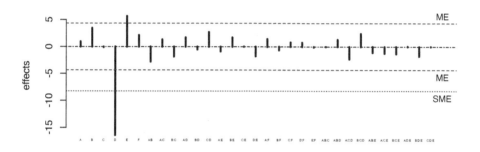

FIGURE 6.10: Lenth Plot for Example 6.5

6.8.1 Effect Aliasing and Design Resolution

Figures 6.9 and 6.10 both indicate that factors E and D may be effective in treating HSV-1. What information could have been obtained if a full 2^6 design had been used?

Factors	Number of effects
Main	$\binom{6}{1} = 6$
2-factor	$\binom{6}{2} = 15$
3-factor	$\binom{6}{3} = 20$
4-factor	$\binom{6}{4} = 15$
5-factor	$\binom{6}{5} = 6$

Factors	Number of effects
6-factor	$\binom{6}{6} = 1$

There are 63 degrees of freedom in a 64 run design. But, 42 are used for estimating three factor interactions or higher. Is it practical to commit two-thirds the degrees of freedom to estimate such effects? According to the effect hierarchy principle, three-factor interactions and higher are not usually important. Thus, using full factorial is wasteful. It's more economical to use a fraction of a full factorial design that allows lower order effects to be estimated.

The 2^{6-1} design in Example 6.5 assigned the factor F to the five-way interaction ABCDE, so this design cannot distinguish between F and ABCDE. The main factor F is said to be **aliased** with the ABCDE interaction.

This aliasing relation is denoted by

$$F = ABCDE \text{ or } I = ABCDEF,$$

where I denotes the column of all $+$'s.

This uses the same mathematical definition as the confounding of a block effect with a factorial effect. Aliasing of the effects is the price for choosing a smaller design.

The 2^{6-1} design has only 31 degrees of freedom for estimating factorial effects, so cannot estimate all 63 factorial effects among the factors A, B, C, D, E, and F.

The equation $I = ABCDEF$ is called the **defining relation** of the 2^{6-1} design. The design is said to have resolution V because the defining relation consists of the "word" ABCDE, which has "length" 5.

Multiplying both sides of $I = ABCDEF$ by column A

$$A = A \times I = A \times ABCDEF = BCDEF,$$

the relation $A = BCDEF$ is obtained. A is aliased with the BCDEF interaction. The aliasing effect for the two-factor interaction AB is obtained by multiplying both sides of $I = ABCDEF$ by AB

$$AB = AB \times I = AB \times ABCDEF = CDEF.$$

The two-factor interaction AB is aliased with the four-factor interaction CDEF. Table 6.11 shows all 31 aliasing relations for the design in Example 6.11.

TABLE 6.11: Aliases for the Design of Example 6.5

Alias
A = BCDEF
B = ACDEF
C = ABDEF
D = ABCEF
E = ABCDF
F = ABCDE
AB = CDEF
AC = BDEF
BC = ADEF
AD = BCEF
BD = ACEF
CD = ABEF
AE = BCDF
BE = ACDF
CE = ABDF
DE = ABCF
AF = BCDE
BF = ACDE
CF = ABDE
DF = ABCE
EF = ABCD
ABC = DEF
ABD = CEF
ACD = BEF
BCD = AEF
ABE = CDF
ACE = BDF
BCE = ADF
ADE = BCF
BDE = ACF

$$CDE = ABF$$

6.8.2 Computation Lab: Fractional Factorial Designs

The design in Example 6.5 has six main effects with aliases: A = BCDEF, B = ACDEF, C = ABDEF, D = ABCEF, E = ABCDF, F = ABCDE. Therefore, the main effects of A, B, C, D, E, and F are estimable only if the aforementioned five-factor interactions are negligible. The other factorial effects have analogous aliasing properties.

The techniques that were covered for analysis of factorial designs are applicable to fractional factorial designs.

The FrF2() function can also be used for generating fractional factorial designs. The design for Example 6.5 is generated using FrF2::FrF2().

```
FrF2::FrF2(
  nruns = 32,
  nfactors = 6,
  factor.names = c("E", "D", "C", "B", "A", "F"),
  randomize = F
)
```

The data frame hsvdat contains the design matrix and dependent variable readout for Example 6.5. The factorial effects can be calculated using linear regression.

```
hsv_mod <- lm(readout ~ A * B * C * D * E * F, data = hsvdat)
2 * coef(hsv_mod)[-1]
```

##	A	B	C	D
##	0.9500	3.4500	-0.0875	-16.4250
##	E	F	A:B	A:C
##	5.6375	2.1250	-2.7625	1.3000
##	B:C	A:D	B:D	C:D
##	-1.8000	1.6875	-0.5125	2.7000
##	A:E	B:E	C:E	D:E
##	-0.8750	1.7000	0.0375	-1.7500
##	A:F	B:F	C:F	D:F
##	1.4125	-0.6125	0.8000	0.7875
##	E:F	A:B:C	A:B:D	A:C:D

```
##      -0.0750       -0.0125        1.3000       -2.2875
##        B:C:D         A:B:E         A:C:E         B:C:E
##       2.4125       -1.0625       -1.2250       -1.3000
##        A:D:E         B:D:E         C:D:E         A:B:F
##       0.0625       -1.7625        0.0250            NA
##        A:C:F         B:C:F         A:D:F         B:D:F
##           NA            NA            NA            NA
##        C:D:F         A:E:F         B:E:F         C:E:F
##           NA            NA            NA            NA
##        D:E:F       A:B:C:D       A:B:C:E       A:B:D:E
##           NA            NA            NA            NA
##      A:C:D:E       B:C:D:E       A:B:C:F       A:B:D:F
##           NA            NA            NA            NA
##      A:C:D:F       B:C:D:F       A:B:E:F       A:C:E:F
##           NA            NA            NA            NA
##      B:C:E:F       A:D:E:F       B:D:E:F       C:D:E:F
##           NA            NA            NA            NA
##    A:B:C:D:E     A:B:C:D:F     A:B:C:E:F     A:B:D:E:F
##           NA            NA            NA            NA
##    A:C:D:E:F   B:C:D:E:F   A:B:C:D:E:F
##           NA            NA            NA
```

The aliases can be computed using the `FrF2::aliases()`.

```
FrF2::aliases(hsv_mod)
```

```
##
##   A = B:C:D:E:F
##   B = A:C:D:E:F
##   C = A:B:D:E:F
##   D = A:B:C:E:F
##   E = A:B:C:D:F
##   F = A:B:C:D:E
##   A:B = C:D:E:F
##   A:C = B:D:E:F
##   B:C = A:D:E:F
##   A:D = B:C:E:F
##   B:D = A:C:E:F
##   C:D = A:B:E:F
##   A:E = B:C:D:F
##   B:E = A:C:D:F
##   C:E = A:B:D:F
##   D:E = A:B:C:F
```

```
##   A:F = B:C:D:E
##   B:F = A:C:D:E
##   C:F = A:B:D:E
##   D:F = A:B:C:E
##   E:F = A:B:C:D
##   A:B:C = D:E:F
##   A:B:D = C:E:F
##   A:C:D = B:E:F
##   B:C:D = A:E:F
##   A:B:E = C:D:F
##   A:C:E = B:D:F
##   B:C:E = A:D:F
##   A:D:E = B:C:F
##   B:D:E = A:C:F
##   C:D:E = A:B:F
```

6.9   Exercises

Exercise 6.1. Consider a $4^2 \times 3^2 \times 2$ factorial design.

a. How many factors are included in this design?

b. How many levels are included in each factor?

c. How many experimental conditions, or runs, are included in this design?

Exercise 6.2. Consider a 2^2 factorial experiment with factors A and B. Show that $INT(A, B) = INT(B, A)$. That is, the interaction is symmetric in B and A.

Exercise 6.3. Consider a 2^3 experiment with factors A, B, and C. Show that

$$INT(A, B, C) = \frac{1}{2}[INT(A, C|B+) - INT(A, C|B-)]$$
$$= \frac{1}{2}[INT(C, B|A+) - INT(C, B|A-)].$$

Exercise 6.4. Example 6.1 used

```
sum(wtlossdat["B"]*wtlossdat["y"])/4
```

to compute the main effect of C.

a. Explain why the espression is divided by 4?

b. Write an R function to calculate the main effects using the design matrix.

c. Write an R function to calculate the interaction effects using the design matrix.

Exercise 6.5. Show that the variance of the ith run in a 2^3 design with two replications is

$$\frac{(y_{i1} - y_{i2})^2}{2},$$

where y_{ij} is j^{th} observation in i^{th} experimental run.

Exercise 6.6. Show that the variance of a factorial effect in a 2^k design with m replications is

$$\frac{4\sigma^2}{N},$$

where $N = m \cdot 2^k$ and σ^2 is the variance of each outcome.

Exercise 6.7. Assume that y_{ij} are i.i.d and follow a normal distribution with variance σ^2. Under the null hypothesis that a factorial effect 0,

a. Show that

$$\frac{\bar{y}_+ - \bar{y}_-}{s/2} \sim t_{2^k(m-1)}$$

in a 2^k design with m replications.

b. Suppose that y_{ij} are not normal. Why is it still reasonable to assume that the ratio of the factorial effect to its standard error follows a normal distribution?

Exercise 6.8. Use R to calculate the three-way interaction effects from the experiment in Example 6.2. Interpret these interaction terms.

Exercise 6.9. This exercise is based on Section 5.2 of Box et al. [2005]. An experiment for optimizing response yield (y) in a chemical plant operation was conducted. The three factors considered are temperature (T), concentration (C), and catalyst type (K). The order of the runs (**run**) was randomized. The levels for the factors are listed below.

Factor	Level
T. Temperature (°C) 1	60 (-1), 180(+1)
C. Concentration (%)	20 (-1), 40 (+1)
K. Catalyst	A (-1), B (+1)

The data are available in the **chemplant** data frame.

```
glimpse(chemplant, n = 3)
```

```
## Rows: 16
## Columns: 5
## $ run <int> 6, 2, 1, 5, 8, 9, 3, 7, 13, 4, 16, 10, 12~
## $ T   <int> -1, 1, -1, 1, -1, 1, -1, 1, -1, 1, -1, 1,~
## $ C   <int> -1, -1, 1, 1, -1, -1, 1, 1, -1, -1, 1, 1,~
## $ K   <int> -1, -1, -1, -1, 1, 1, 1, 1, -1, -1, -1, -~
## $ y   <int> 59, 74, 50, 69, 50, 81, 46, 79, 61, 70, 5~
```

a. State the design.

b. Use R to calculate all the factorial effects.

c. Which factors have an important effect on yield? Explain how you arrived at your answer.

Exercise 6.10. Interpret the intercept term in the linear regression model represented in Example 6.1.

Exercise 6.11. Suppose that a 2^2 factorial design studying factors A and B was conducted. An investigator fits the model

$$y_i = \beta_0 x_{i0} + \beta_1 x_{i1} + \beta_2 x_{i2} + \beta_3 x_{i3} + \epsilon_i,$$

$i, \ldots, n$, to estimate the factorial effects of the study.

a. What is the value of n? Define $x_{ik}, k = 0, 1, 2, 3$.

b. Derive the least-squares estimates of $\beta_k, k = 0, 1, 2, 3$.

c. Show that the least-squares estimates of A, B, AB, are one-half the factorial effects.

Exercise 6.12. Suppose that linear regression is used to estimate factorial effects for a 2^k design by doubling the estimated regression coefficients.

a. When is it possible to estimate the standard error of these regression coefficients?

b. If the p-values for the regression coefficients are able to be calculated, then is it necessary to transform these p-values to correspond to p-values for factorial effects? Explain.

Exercise 6.13. Suppose Y has a half-normal distribution with variance $\sigma^2 = 1$. Let y_p be the pth quantile of Y such that $p = F(y_p)$ and Φ be the CDF of the standard normal distribution.

a. Show that $F(y_p) = 2\Phi(y_p) - 1$.

b. Show that the pth quantile of a half-normal is $y_p = \Phi^{-1}(p/2 + 1/2)$.

Exercise 6.14. In Section 6.5.2 Computation Lab: Normal Plots in Unreplicated Factorial Designs, the factorial effects were sorted based on their magnitudes before plotting the normal quantile plot shown in Figure 6.7.

a. Create the normal quantile plot without sorting the factorial effects. What is wrong with the plot?

b. Instead of using **geom_qq()** and **geom_qq_line()**, construct a normal quantile plot using **geom_point()** and add an appropriate straight line to the plot that connects the first and third quantiles of the theoretical and sample distributions.

c. Suppose the absolute values of factorial effects are used in a normal quantile plot. Will the result be an interpretable plot to assess the significance of factorial effects in an unreplicated design? Explain.

Exercise 6.15. Reconstruct the half-normal plot shown in Figure 6.8 using **geom_qq()** and **geom_qq_line()**.

Exercise 6.16. The planning matrix from a 2^{k-p} design is below. The four factors investigated are A, B, C, and D.

TABLE 6.13: Planning Matrix for Exercise 6.16

A	B	C	D
−	−	−	−
+	−	−	+
−	+	−	+
+	+	−	−
−	−	+	+
+	−	+	−
−	+	+	−
+	+	+	+

a. What are the values of k and p?

b. What is the defining relation? What is the resolution of this design?

c. What are the aliasing relations?

d. Use `FrF2::FrF2()` to generate this design.

Exercise 6.17. Use the `FrF2()::FrF2()` function to generate the design matrix in the standard order for a 2^3 design with blocks generated using the three-way interaction.

Exercise 6.18. An investigator is considering two blocking schemes for a 2^4 design with 4 blocks. The two schemes are listed below.

- Scheme 1: $B_1 = 134$, $B_2 = 234$
- Scheme 2: $B_1 = 12$, $B_2 = 13$, $B_3 = 1234$

a. Which runs are assigned to which blocks in each scheme?

b. For each scheme, list all interactions confounded with block effects.

c. Which blocking scheme is preferred? Explain why.

Exercise 6.19. This exercise is based on Section 5.1 of Wu and Hamada [2011]. An experiment to improve a heat treatment process on truck leaf springs was conducted. The heat treatment that forms the camber in leaf springs consists of heating in a high temperature furnace, processing by forming a machine, and quenching in an oil bath. The free height of an unloaded spring has a target value of 8 in. The goal of the experiment is to make the variation about the target as small as possible.

The five factors studied in this experiment are shown below. The data are available in the data frame `leafspring`, where y1, y2, y3 are three replications of the free height measurements.

Factor	Level
B. Temperature	1840 (-), 1880 (+)
C. Heating time	23 (-), 25 (+)
D. Transfer time	10 (-), 12 (+)
E. Hold down time	2 (-), 3 (+)
Q. Quench oil temperature	130-150 (-), 150-170 (+)

```
glimpse(leafspring, n = 3)
```

```
## Rows: 16
## Columns: 8
## $ B  <chr> "-", "+", "-", "+", "-", "+", "-", "+", "-~
## $ C  <chr> "+", "+", "-", "-", "+", "+", "-", "-", "+~
## $ D  <chr> "+", "+", "+", "+", "-", "-", "-", "-", "+~
## $ E  <chr> "-", "+", "+", "-", "+", "-", "-", "+", "-~
## $ Q  <chr> "-", "-", "-", "-", "-", "-", "-", "-", "+~
## $ y1 <dbl> 7.78, 8.15, 7.50, 7.59, 7.94, 7.69, 7.56, ~
## $ y2 <dbl> 7.78, 8.18, 7.56, 7.56, 8.00, 8.09, 7.62, ~
## $ y3 <dbl> 7.81, 7.88, 7.50, 7.75, 7.88, 8.06, 7.44, ~
```

Answer the questions below.

a. State the design.

b. What is the defining relation? Which of the factorial effects are aliased?

c. The goal of this experiment was to investigate which factors minimize variation of free height. Use the sample variance for each run as the dependent variable to estimate the factorial effects.

d. Which factorial effects have an important effect on free height variation?

Exercise 6.20. A baker (who is also a scientist) designs an experiment to study the effects of different amounts of butter, sugar, and baking powder on the taste of his chocolate chip cookies.

Factor	Amount
B. Butter	10g (-), 15g (+)
S. Sugar	1/2 cup (-), 3/4 cup (+)
P. Baking Powder	1/2 teaspoon (-), 1 teaspoon (+)

The response variable is taste measured on a scale of 1 (poor) to 10 (excellent). He first designs a full factorial experiment based on the three factors butter, sugar, and baking powder. But, before running the experiment the baker decides to add another factor of baking time (T)—12 minutes vs. 16 minutes. However, he cannot afford to conduct more than 8 runs since each run requires time to prepare a different batch of cookie dough, bake the cookies, and measure taste. So, he decides to use the three factor interaction between butter, sugar, and baking powder to assign whether baking time will be 12 minutes (-) or 16 minutes (+) in each of the 8 runs.

The data frame `cookies` contains the data for this experiment, but does not contain a column for baking time.

`glimpse(cookies)`

```
## Rows: 8
## Columns: 5
## $ Run    <dbl> 1, 2, 3, 4, 5, 6, 7, 8
## $ Butter <dbl> -1, 1, -1, 1, -1, 1, -1, 1
## $ Sugar  <dbl> -1, -1, 1, 1, -1, -1, 1, 1
## $ Powder <dbl> -1, -1, -1, -1, 1, 1, 1, 1
## $ Taste  <dbl> 3, 3, 10, 2, 6, 5, 4, 6
```

a. Add a column for baking time (T) to `cookies`.

b. State the design.

c. Which factors have an important effect on taste?

d. Are any of the factors aliased? If yes, then state the defining relation and aliasing effects.

e. The cookie batches were baked without randomizing the order of the runs. What impact, if any, would this have on the results? Explain.

f. In order to speed up the experiment, the baker was contemplating using his neighbour's oven to bake runs 1, 4, 6, and 7. What impact, if any, could this have on the results? Explain.

7

Causal Inference

7.1 Introduction: The Fundamental Problem of Causal Inference

Consider the following *hypothetical* scenario. Kayla, a high school senior sprinter, is contemplating whether or not to run the Canadian National High School 100 metre final race in new spikes (Cheetah 5000—a new high tech racing spike), or her usual brand of racing spikes. Her goal is to obtain a time below 12s (seconds).

This scenario can be expressed as a scientific question using terms from previous chapters. There are two **treatment levels**, wearing the new spikes or wearing her usual spike shoes during the race. If Kayla *wears the new shoes*, her time may be less than 12s, or it may be greater; we denote this *potential* race outcome, which can be either "Fast" (less than 12s) or "Slow" (greater than or equal to 12s), by $Y(1)$. Similarly, if Kayla wears her standard racing spikes (i.e., not the Cheetah 5000), her race time may or may not be below 12s; we denote this *potential* race outcome by $Y(0)$, which also can be either "Fast" or "Slow". Table 7.1 shows the two **potential outcomes**, $Y(1)$ and $Y(0)$, one for each level of the treatment. The causal effect of the treatment involves the comparison of these two potential outcomes. In this scenario each potential outcome can take on only two values, the experimental unit level (i.e., for Kayla) **causal effect** – the comparison of these two outcomes for the same experimental unit – involves one of four (two by two) possibilities (Table 7.1).

TABLE 7.1: Potential Outcomes of Kayla's Race

Wear Cheetah 5000s	Potential Race Outcomes	
Yes (1)	Y(1) = Fast	Y(1) = Slow
No (0)	Y(0) = Fast	Y(0) = Slow

There are two important aspects of this definition of a causal effect.

DOI: 10.1201/9781003033691-7

1. The definition of the causal effect depends on the potential outcomes, but it does not depend on which outcome is actually observed. Specifically, whether Kayla wears the Cheetah 5000 or standard spikes does not affect the definition of the causal effect.

2. The causal effect is the comparison of potential outcomes, for Kayla (the same experimental unit), at the Canadian National High School 100m (metre) final race (i.e., a certain moment in time) after putting the racing spikes on her feet and running the race (i.e., post-treatment). In particular, the causal effect is *not* defined in terms of comparisons of outcomes at different races, as in a comparison of Kayla's race results after running one 100m race wearing the Cheetah 5000 and another 100m race wearing standard racing spikes.

The **fundamental problem of causal inference** is the problem that at most one of the potential outcomes can be realized and thus observed [Holland, 1986]. If the action Kayla takes is racing the Canadian National race, in the Cheetah 5000, you observe $Y(1)$ and will never know the value of $Y(0)$ because you cannot go back in time. Similarly, if her action is wearing standard racing spikes, you observe $Y(0)$ but cannot know the value of $Y(1)$. In general, therefore, even though the unit-level causal effect (the comparison of the two potential outcomes) may be well defined, by definition we cannot learn its value from just the single realized potential outcome. The outcomes that would be observed under control (standard spikes) and treatment (Cheetah 5000) conditions are often called **counterfactuals or potential outcomes**.

If Kayla wears the Cheetah 5000 for the Canadian Nationals then $Y(1)$ is observed and $Y(0)$ is the unobserved counterfactual outcome—it represents what would have happened to Kayla if she wears standard spikes. Conversely, if Kayla wears standard spikes for the Canadian National race then $Y(0)$ is observed and $Y(1)$ is counterfactual. In either case, a simple treatment effect for Kayla can be defined as $Y(1) - Y(0)$.

Example 7.1 (Fundemental Problem of Causal Inference). Table 7.2 shows hypothetical data for an experiment with 100 units (200 potential outcomes). The table shows the data that is required to determine causal effects for each unit in the data set — that is, it includes both potential outcomes for each person. Let T_i be an indicator of one of the two treatment levels. $Y_i(0)$ be the potential outcome for unit i when $T_i = 0$ and $Y_i(1)$ be the potential outcome for unit i when $T_i = 1$.

The **fundamental problem of causal inference** is that at most one of these two potential outcomes, $Y_i(0)$ and $Y_i(1)$, can be observed for each unit i. Table 7.3 displays the data that can actually be observed. The $Y_i(1)$ values are "missing" for those who received treatment $T = 0$, and the $Y_i(0)$ values are "missing" for those who received $T = 1$.

TABLE 7.2: Observed Data with Potential Outcomes

Unit	T_i	$Y_i(0)$	$Y_i(1)$	$Y_i(0) - Y_i(1)$
1	0	69	75	-6
2	1	80	76	4
3	1	71	69	1
$\vdots$	$\vdots$	$\vdots$	$\vdots$	$\vdots$
100	0	81	78	3

TABLE 7.3: Observed Data without Potential Outcomes

Unit	T_i	$Y_i(0)$	$Y_i(1)$	$Y_i(0) - Y_i(1)$
1	0	**69**	?	?
2	1	?	**76**	?
3	1	?	**69**	?
$\vdots$	$\vdots$	$\vdots$	$\vdots$	$\vdots$
100	0	**81**	?	?

We cannot observe both what happens to an individual after taking the treatment (at a particular point in time) and what happens to that same individual after not taking the treatment (at the same point in time). Thus we can never measure a causal effect directly.

7.1.1 Stable Unit Treatment Value Assumption (SUTVA)

The assumption. The potential outcomes for any unit do not vary with the treatments assigned to other units, and, for each unit, there are no different forms or versions of each treatment level, which lead to different potential outcomes.

The stable unit treatment value assumption involves assuming that treatments applied to one unit do not affect the outcome for another unit. For example, if Adam and Oliver are in different locations and have no contact with each other, it would appear reasonable to assume that if Oliver takes an aspirin for his headache, then his behaviour has no effect on the status of Adam's headache. This assumption might not hold if Adam and Oliver are in the same location, and Adam's behavior affects Oliver's behaviour. SUTVA incorporates the idea that Adam and Oliver do not *interfere* with one another and the idea that for each unit there is only a single version of each treatment level (e.g., there is only one version of aspirin used) [Imbens and Rubin, 2015].

The causal effect of aspirin on headaches can be estimated if we are able to exclude the possibility that your taking or not taking aspirin has any effect on my headache, and that the aspirin tablets available to me are of different strengths. These are assumptions that rely on previously acquired knowledge of the subject matter for their justification. Causal inference is generally impossible without such assumptions, so it's important to be explicit about their content and their justifications [Imbens and Rubin, 2015].

7.2 Treatment Assignment

A **treatment assignment vector** records the treatment that each experimental unit is assigned to receive. Let N be the number of experimental units, and T_i a treatment indicator for unit i

$$T_i = \begin{cases} 1 & \text{if subject i assigned treatment A} \\ 0 & \text{if subject i assigned treatment B.} \end{cases}$$

If $N = 2$, then the possible treatment assignment vectors $(t_1, t_2)'$ are $(1, 0)'$, $(0, 1)'$, $(1, 1)'$, and $(0, 0)'$. For example, the first treatment assignment vector $(1, 0)'$ means that the first experimental unit receives treatment A, and the second treatment B.

Example 7.2 (Dr. Perfect). The following example is adapted from Imbens and Rubin [2015]. Suppose Dr. Perfect, an AI algorithm, can, with perfect accuracy, predict survival after treatment $T = 1$ or no treatment $T = 0$ for a certain disease. Let $Y_i(0)$ and $Y_i(1)$ be years of survival (potential outcomes) for the ith patient under no treatment and treatment, and T_i treatment for the ith patient.

TABLE 7.4: Dr. Perfect Treatment Comparison with Potential Outcomes

Unit	$Y_i(0)$	$Y_i(1)$	$Y_i(1) - Y_i(0)$
patient #1	1	7	6
patient #2	6	5	-1
patient #3	1	5	4
patient #4	8	7	-1
Average	4	6	2

A patient's doctor will prescribe treatment only if their survival is greater compared to not receiving treatment.

Patients receiving treatment on average live two years longer. Patients 1 and 3 will receive treatment and patients 2 and 4 will not receive treatment (Table 7.4).

TABLE 7.5: Dr. Perfect Treatment Comparison Observed Outcomes

Unit	T_i	Y_i^{obs}	$Y_i(0)$	$Y_i(1)$
patient #1	1	7	?	7
patient #2	0	6	6	?
patient #3	1	5	?	5
patient #4	0	8	8	?
Average			7	6

where

$$T_i = \begin{cases} 1 & \text{if } Y_i(1) > Y_i(0) \\ 0 & \text{if } Y_i(1) \leq Y_i(0) \end{cases} \tag{7.1}$$

Observed survival can be defined as: $Y_i^{obs} = T_i Y_i(1) + (1 - T_i)Y_i(0)$, and the *observed* data (Table 7.5) tells us that patients receiving treatment live, on average, one year longer. This shows that we can reach invalid conclusions if we look at the observed values of potential outcomes without considering how the treatments were assigned.

In this case, treatment assignment depends on potential outcomes, so the probability of receiving treatment depends on the missing potential outcome.

If we were to draw a conclusion based on the observed sample means, then no treatment adds one year of life, but the truth is that on average treatment adds two years of life. What happened? Let's start by looking at the mean difference in all possible treatment assignments, or the randomization distribution of the mean difference.

Let $\mathbf{T} = (t_1, t_2, t_3, t_4)' = (1, 0, 1, 0)'$ be the treatment assignment vector (see 3.2.1) in Example 7.2. There are $\binom{4}{2}$ possible treatment assignment vectors where two patients are treated and two patients are not treated. Table 7.6 shows the mean difference under all possible treatment assignments—the notation 1010 means the first and third patients receive treatment (since there is a 1 in the first and third place) and the second and fourth did not receive treatment (since there is a 0 in the first and third place). We see that the average mean difference is 0 which equals the "true" difference between treatment and no treatment.

In contrast, Dr. Perfect's assignment depends on the potential outcomes, and we observed an extreme mean difference, but if treatment assignment had been random, then it would not depend on the potential outcomes, and on average the inference would have been correct. We can operationalize randomly assigning treatments by randomly selecting a card for each patient where two cards are labeled "treat" and two labeled "no treat". Then

$$T_i = \begin{cases} 1 & \text{if "treat"} \\ 0 & \text{if "no treat"} \end{cases} \tag{7.2}$$

Definition 7.1 (Assignment Mechanism). The probability that a particular treatment assignment will occur, $P(\mathbf{T}|X, Y(0), Y(1))$, is called the assignment mechanism. Note that this is **not** the probability a particular experimental unit will receive treatment, but is the probability of a value for the full treatment assignment.

Table 7.6 shows all the possible treatment assignments for Example 7.2 using (7.2). In this case **T** belongs to

$$\mathbf{W} = \left\{ \begin{pmatrix} 1 \\ 1 \\ 0 \\ 0 \end{pmatrix}, \begin{pmatrix} 1 \\ 0 \\ 1 \\ 0 \end{pmatrix}, \begin{pmatrix} 1 \\ 0 \\ 0 \\ 1 \end{pmatrix}, \begin{pmatrix} 0 \\ 1 \\ 1 \\ 0 \end{pmatrix}, \begin{pmatrix} 0 \\ 1 \\ 0 \\ 1 \end{pmatrix}, \begin{pmatrix} 0 \\ 0 \\ 1 \\ 1 \end{pmatrix} \right\},$$

since two units are treated and two units are not treated.

$$P(\mathbf{T}|X, Y(0), Y(1)) = \begin{cases} 1/\binom{4}{2} & \text{if } \mathbf{T} \in \mathbf{W} \\ 0 & \text{if } \mathbf{T} \notin \mathbf{W}. \end{cases}$$

Note that this probability is defined on $\{(w_1, w_2, w_3, w_4)' | w_i \in \{0, 1\}\}$.

Definition 7.2 (Unconfounded Assignment Mechanism). An assignment mechanism is **unconfounded** if $P(\mathbf{T}|X, Y(0), Y(1)) = P(\mathbf{T}|X)$.

T_i in (7.2) is a random variable whose distribution is independent of the potential outcomes $P(T_i = 1|Y_i(0), Y_i(1)) = P(T_i = 1) = 1/2$. Compare this with the distribution of T_i in (7.1), where the marginal distribution of T_i is $P(T_i) = 1/2$, so T_i is not independent of $Y_i(0), Y_i(1)$. In this case, there are no covariates, so we omit X.

If the treatment assignment in (7.2) is used, then we can **ignore** the treatment indicator and use the observed values for causal inference, since treatment assignment is **unconfounded** (i.e., not confounded) with potential outcomes.

Definition 7.3 (Propensity Score). The propensity score is the probability, $P(T_i = 1|X_i)$, that a particular experimental unit receives treatment.

The propensity score in Example 7.2 using either the treatment assignments (7.1) or (7.2) is $P(T_i) = 1/2$.

TABLE 7.6: All Possible Treatment Assignments: Dr. Perfect (Cont'd.)

Treatment Assignment	Mean Difference
1100	1.5
1010	-1.0
1001	3.5
0110	0.5
0101	5.0
0011	2.5

Example 7.3 (Lord's Paradox). Wainer and Brown [2004] describes an investigation related to answering "how much does group membership [race] matter in measuring the effect of medical school?" This is investigated by examining "an individual's rank among an incoming class of medical students based on the Medical College Admissions Test (MCAT), and then examining rank after receiving a major portion of her medical education based on the first US medical licensing exam (USMLE- Step 1). If rank does not change, we could conclude that the effect of medical school was the same for that individual as it was for the typical medical student."

Following Lord [1967], we frame the problem as two statisticians analyzing the data. Statistician 1 compares the average change in ranks with one group compared to another, and finds that white examinees' ranks improved, about 19 places, and black examinees declined by 19 places, for a differential effect of 38. Statistician 2 uses pre-medical school ranks, measured by the MCAT, as a covariate and examines the differences between groups' average medical school rank after adjusting for the pre-medical school rank, and finds that white examinees' ranks improved by about 9 places, and black examinees declined by about 9, for a differential effect of 18. Which statistician is correct?

The experimental units are medical students before entering medical school. The treatment is medical school and the control is *unknown*. The assignment mechanism is unconfounded, but all students went to medical school so we don't observe an outcome, $Y(0)$ under the control condition. This means that the propensity score for each student, $P(T_i = 1|X_i) = 1$.

Which statistician is correct? It's possible to make either statistician correct depending on what we are willing to assume. Statistician 1 must assume that however the control treatment is defined (e.g., a different medical school) they would have the same rank in the incoming class (i.e., MCAT score). Statistician 2 must assume that a student's rank in medical school, under the *unknown* control condition, is a linear function of their pre-medical school rank, and the same linear function must apply to all the students.

Both statisticians can be made correct even though they come to different conclusions, so neither is correct. Everyone went to medical school and a control group was not specified, so the data has no information about the effect of skin colour on rank in medical school. In order to draw causal conclusions, we need a probablistic assignment mechanism, $0 < P(T_i = 1|X_i) < 1$, in addition to an assignment mechanism that is unconfounded.

Definition 7.4 (Probabilistic Assignment). An assignment mechanism is **probabilistic** if $0 < P(\mathbf{T}|X, Y(0), Y(1)) < 1$

Example 7.4 (Assigning Treatments Using Coin Tossing). Suppose an investigator randomly assigns N experimental units to treatment or control by tossing a coin, where the probability of a head is p. Each person's propensity score is p. The assignment mechanism is $P(\mathbf{T}|Y(0), Y(1), X) = p^N$.

Example 7.5 (Assignment Mechanism—Randomized Block Design). Consider a randomized design where the researcher decides to block on a binary covariate X, and each block is to consist of 50 experimental units, where 25 units receive treatment and 25 receive control. The propensity score for each unit is 0.5, and the assignment mechanism is $P(\mathbf{T}_1|Y(0), Y(1), X = 1) = P(\mathbf{T}_0|Y(0), Y(1), X = 0) = \binom{50}{25}^{-1}$. The overall assignment mechanism is $P((\mathbf{T}_0, \mathbf{T}_1)|Y(0), Y(1), X) = \binom{50}{25}^{-2}$

Example 7.6 (Assignment Mechanism—Randomized Pair Design). A randomized pair design is a block design where the block size is 2. Within each pair, one unit is randomly assigned to treatment and the other to control, so each subject has a propensity score of p. If we have N pairs, then the assignment mechanism is $P(\mathbf{T}|Y(0), Y(1), X) = p^N$.

7.2.1 Computation Lab: Treatment Assignment

The treatment assignments and mean differences for each treatment assignment in Example 7.2 can be computed in R.

```
aidat <- tibble(Y0 = c(1, 6, 1, 8), Y1 = c(7, 5, 5, 7))
trt_assignments <- combn(1:4, 2)

N <- choose(4, 2)
delta <- numeric(N)

for (i in 1:N) {
  delta[i] <- mean(aidat$Y1[trt_assignments[, i]]) -
```

```
    mean(aidat$Y0[-trt_assignments[, i]])
}

mean(delta)
```

```
## [1] 2
```

This is very similar to the randomization distribution of Chapter 3 except that we are using a hypothetical example where we know all the potential outcomes for each subject. So, in this case there is no need to *exchange* treated for untreated subjects.

The confounded assignment mechanism in Example 7.2 using (7.1) is computed below. `trt_assignments` depends on `Y1` and `Y0`.

```
aidat <- tibble(Y0 = c(1, 6, 1, 8), Y1 = c(7, 5, 5, 7))

trt_assignments <- ifelse(aidat$Y0 > aidat$Y1, 0, 1)

Yobs <- ifelse(trt_assignments == 1, aidat$Y1, aidat$Y0)

mean(Yobs[trt_assignments == 1]) -
  mean(Yobs[trt_assignments == 0])
```

```
## [1] -1
```

Example 7.7 (Add Randomness to Dr. Perfect). In this example we will add some randomness to the treatment assignment (7.1). If $Y_i(1) > Y_i(0)$, then patient i will receive treatment if a biased coin is heads, where the probability of heads is 0.8; and if $Y_i(1) \leq Y_i(0)$, then patient i will receive treatment if a biased coin is heads, where the probability of heads is 0.3.

This can be implemented using the code below. `rbinom(1, 1, p)` simulates a coin flip, where the probability of head is `p`. The observed data `Yobs` requires a bit of extra attention since it's possible that all patients might be treated or not treated. Calculation of the mean difference also requires some thought, since we must account for the cases when all subjects are treated or all subjects are not treated.

```
set.seed(21)
aidat <- tibble(Y0 = c(1, 6, 1, 8), Y1 = c(7, 5, 5, 7))
trt_assignments <- mapply(function(y0, y1)
```

```
  if_else(y0 > y1,
          rbinom(1, 1, .3),
          rbinom(1, 1, .8)),
  aidat$Y0, aidat$Y1)
Nobs <- 4

for (i in 1:Nobs) {
  if (trt_assignments[i] == 1) {
    Yobs[i] <- aidat$Y1[i]
  } else
    if (trt_assignments[i] == 0) {
      Yobs[i] <- aidat$Y0[i]
    }
}

if (sum(trt_assignments) > 0 & sum(trt_assignments) < Nobs) {
  mean(Yobs[trt_assignments == 1]) -
    mean(Yobs[trt_assignments == 0])
} else
  if (sum(trt_assignments) == 0) {
    mean(Yobs[trt_assignments == 0])
  } else
    if (sum(trt_assignments) == Nobs) {
      mean(Yobs[trt_assignments == 1])
    }
```

```
## [1] 2
```

7.3   Causal Effects and Randomized Experiments

Classical randomized experiments such as completely randomized designs, randomized paired comparisons, and randomized block designs all have unconfounded, probabilistic assignment mechanisms. Examples 7.3 and 7.2 illustrate that for causal inference a probabilistic, unconfounded assignment mechanism defines these classical randomized experiments.

Randomization and experimentation are one approach to dealing with the fundamental problem of causal inference. Outcomes are observed on a sample of units to learn about the distribution of outcomes in the population. The fundamental problem states that we cannot compare treatment and control outcomes for the same units, so we try to compare them on similar units. Similarity can be attained by using randomization to decide which units are assigned to the treatment group and which units are assigned to the control group. Other approaches to dealing with the problem are statistical adjustment via regression modelling and finding close substitutes to the units that did not receive treatment.

In most practical situations it's impossible to estimate individual-level causal effects, but we can design studies to estimate treatment effects such as:

$$\text{average causal effect} = \bar{Y}(1) - \bar{Y}(0),$$
$$\text{median causal effect} = \text{Median}\{\mathbf{Y}(1) - \mathbf{Y}(0)\},$$

where $\bar{Y}(k) = \sum_{i=1}^{N_k} Y_i(k)/N_1$, $k = 1, 2$ and $\mathbf{Y}(k) = (Y_1(k), \ldots, Y_{N_k}(k))'$, $k = 1, 2$.

The control group is a group of units that could have ended up in the treatment group, but due to chance they just happened not to get the treatment. Therefore, on average, their outcomes represent what would have happened to the treated units had they not been treated; similarly, the treatment group outcomes represent what might have happened to the control group had they been treated.

If N_0 units are selected at random from the population and given the control, and N_1 other units are randomly selected and given the treatment, then the observed sample averages of Y for the treated and control units can be used to estimate the corresponding population quantities, $\bar{Y}(1)$ and $\bar{Y}(0)$, with their difference estimating the average treatment effect (and with standard error $\sqrt{S_0^2/N_0 + S_1^2/N_1}$). This works because the $Y_i(0)$ for the control group are a random sample of the values in the entire population. Similarly, the $Y_i(1)$ for the treatment group are a random sample of the $Y_i(1)$'s in the population.

Equivalently, if we select $N_0 + N_1$ units at random from the population, and then randomly assign N_0 of them to the control and N_1 to the treatment, we can think of each of the sample groups as representing the corresponding population of control or treated units. Therefore, the control group mean can act as a counterfactual for the treatment group (and vice versa).

In medical studies such as clinical trials, it is common to select $N_0 + N_1$ units nonrandomly from the population, but then the treatment is assigned at random within this sample. Causal inferences are still justified, but the inferences no longer generalize to the entire population (Gelman and Hill [2006]).

7.4   Causal Effects and Observational Studies

7.4.1   What is an Observational Study?

Randomized experiments are currently viewed as the most credible basis for determining effects of treatments or cause and effect relationships. Health Canada, the U.S. Food and Drug Administration, European Medicines Agency, and other regulatory agencies all rely on randomized experiments in their approval processes for pharmaceutical treatments.

When randomized experimentation is unethical or infeasible, and a researcher aims to study the effects caused by treatments, then the study is called **observational**.

Definition 7.5 (Observational Study). An **observational study** is an empirical study of the effects caused by treatments where the assignment mechanism is unknown.

This defines an observational study as a randomized study whose assignment mechanism is unknown and so is not under control of the researcher. In other words, the researcher doesn't know how treatments were assigned to units. This is further discussed in Imbens and Rubin [2015].

Example 7.8. Doll and Hill [1950] conducted a study to address an increase in the number of deaths attributed to lung cancer. They identified 709 lung cancer patients in twenty London hospitals and enrolled a comparison group of 709 non-cancer patients who had been admitted to the same hospitals. The cancer patients and non-cancer patients were matched by age, gender, and hospital. All subjects were interviewed about their past smoking habits and other exposures using a questionnaire.

The "treatment" in this study is smoking and the outcome is lung cancer. A randomized study to investigate the effects of smoking on lung cancer is neither feasible nor ethical, but it is possible to design a study where the assignment mechanism is unknown. In this study, patients with lung cancer were matched to patients without lung cancer to form two comparable groups without using randomization. This means that the groups with and without cancer will have similar distributions of age, sex, and hospital, but there may be covariates not included in the matching that are associated with smoking and cancer and not recorded in the study. Randomization is supposed to ensure that these unmeasured confounders are balanced between the groups.

7.4.2 Designing an Observational Study

Good observational studies are designed [Rubin, 2007]. It's helpful to *conceive* an observational study as a *hypothetical* randomized experiment with a lost assignment mechanism and rule for propensity scores, whose values will try to be reconstructed. This conceptualization is meant to encourage investigators to envision why experimental units were treated or not treated.

In randomized experiments, the design phase occurs prior to examining the data, and this should also occur in observational studies. The main part of the design stage of an observational study is to assess the degree of balance in the covariate distributions between treated and control units, which involves comparing the distributions of covariates in the treated and control samples. Randomized studies usually generate covariate distributions that are similar in the treated and control groups. When the distributions are similar the covariates are sometimes called *balanced* between the treated and control units, and unbalanced if the distributions are not similar.

7.5 Epidemiologic Follow-up Study

Example 7.9 (Does where you live affect your weight?). The NHANES I Epidemiologic Follow-up Study (NHEFS) [CDC, 2021] is a national longitudinal study that was jointly initiated by the National Center for Health Statistics and the National Institute on Aging in collaboration with other agencies of the United States Public Health Service. The NHEFS was designed to investigate the relationships between clinical, nutritional, and behavioral factors assessed in the first National Health and Nutrition Examination Survey NHANES I and subsequent morbidity, mortality, and hospital utilization, as well as changes in risk factors, functional limitation, and institutionalization.

The NHEFS cohort includes all persons 25-74 years of age who completed a medical examination at NHANES I in 1971-75 (n = 14,407). It is comprised of a series of follow-up studies, four of which have been conducted to date. The first wave of data collection was conducted for all members of the NHEFS cohort from 1982 through 1984. Continued follow-up of the NHEFS population was conducted in 1986, 1987, and 1992 using the same design and data collection procedures developed in the 1982-1984 NHEFS, with the exception that a 30-minute computer-assisted telephone interview was administered rather than a personal interview, and no physical measurements were taken.

What is the effect of living in a city compared to city suburbs on weight gain? The first and fourth waves of the NHEFS provide an opportunity to design an

TABLE 7.7: Distribution of Baseline Variables

Variable	City	Suburbs
Age		
Mean Age 1982 (years)	56.5	52.4
SD Age 1982 (years)	14.2	13.1
Sex		
Male	1074 (34.24%)	793 (36.48%)
Female	2063 (65.76%)	1381 (63.52%)
Smoking 1982		
Yes	933 (51.98%)	657 (49.70%)
No	862 (48.02%)	665 (50.30%)
NA	1342	852
General Health		
Excellent	678 (22.12%)	658 (31.26%)
Very Good	816 (26.62%)	648 (30.78%)
Good	978 (31.91%)	539 (25.61%)
Fair	437 (14.26%)	216 (10.26%)
Poor	156 (5.09%)	44 (2.09%)
NA	72	69
Physical Activity		
Very Active	794 (25.97%)	559 (26.61%)
Moderately Active	1767 (57.80%)	1218 (57.97%)
Quite Inactive	496 (16.23%)	324 (15.42%)
NA	80	73
Marital Status		
Married	2004 (63.90%)	1658 (76.34%)
Widowed	509 (16.23%)	218 (10.04%)
Divorced	339 (10.81%)	158 (7.27%)
Separated	89 (2.84%)	64 (2.95%)
Never Married	195 (6.22%)	74 (3.41%)
NA	1	2
Income		
<= $9,999	760 (26.24%)	284 (13.93%)
$10,000 - $19,999	718 (24.79%)	361 (17.70%)
$20,000 - $34,999	802 (27.69%)	629 (30.85%)
$35,000 - $74,999	537 (18.54%)	660 (32.37%)
>= $75,000	79 (2.73%)	105 (5.15%)
NA	241	135

observational study to evaluate this question.

We define $E(Y|T = 1)$ as the mean weight gain that would have been observed if all individuals in the population had lived in the suburbs before the follow-up visit in 1992 and $E(Y|T = 0)$ as the mean weight gain that would have been observed if all individuals in the population had lived in the city before the follow-up visit in 1992. The **average causal effect** is $E(Y|T = 1) - E(Y|T = 0)$. The observed difference will have a causal interpretation only if people living in the suburbs and city have similar characteristics.

Table 7.7 shows the distribution of several variables in the group of participants that live in suburbs and city. The distributions of sex, smoking, and physical activity are similar or balanced in each group, while the distributions of age, general health, marital status, and income are different.

The next sections will discuss designing an observational study to evaluate whether there is a difference in weight gain between the city and suburbs. The first step is evaluating the distributions of covariates in the group that lived in the suburbs (treated) and the group that lived in the city (control).

7.6 The Propensity Score

Covariates are pre-treatment variables and take the same value for each unit no matter which treatment is applied.

Definition 7.6 (Covariate). A **covariate** is a pre-treatment characteristic of an experimental unit that is not affected by treatment. In many studies, age and sex of a subject would be considered a covariate.

The ith propensity score (Definition 7.3) is $e(\mathbf{x}_i) = P(T_i = 1|\mathbf{x}_i)$, where $\mathbf{x}_i$ is a vector of observed covariates for the ith experimental unit. The ith propensity score is the probability that a unit receives treatment given all the information, recorded as covariates, that is observed before the treatment. In experiments, the propensity scores are known, but in observational studies such as Example 7.9 they can be estimated using statistical models such as logistic regression, where the outcome is the treatment indicator and the predictors are available covariates.

Example 7.10 (Estimating Propensity Scores in NHEFS). The treatment indicator in 7.9 is

$$T_i = \begin{cases} 1 & \text{if } i^{th} \text{ unit lives in suburbs} \\ 0 & \text{if } i^{th} \text{ unit lives in city.} \end{cases}$$

TABLE 7.8: Example Propensity Scores for NHEFS

$\hat{e}(\mathbf{x})$	y	x_1	x_2	x_3	x_4	x_5	x_6	x_7
0.40	2	50	2	1	3	1	3	4
0.43	2	72	1	2	3	2	1	4
0.58	3	39	2	2	2	2	3	2
0.60	3	53	1	1	3	2	1	4

Logistic regression can be used to model $P(T_i|\mathbf{x_i})$, where $\mathbf{x_i}$ is the ith vector of the seven covariates: age; sex; smoking; general health; physical activity; marital status; and income. The ith propensity score can then be estimated by fitting the model

$$\log\left(\frac{P(T_i|\mathbf{x_i})}{1 - P(T|\mathbf{x})}\right) = \sum_{k=1}^{7} \beta_i x_{ik}. \tag{7.3}$$

The estimated propensity scores are:

$$\hat{e}(\mathbf{x}_i) = \frac{\exp\left(\sum_{k=1}^{7} \hat{\beta}_i x_{ik}\right)}{1 + \exp\left(\sum_{k=1}^{7} \hat{\beta}_i x_{ik}\right)},$$

where $\hat{\beta}_i, i = 1, \ldots, 7$, are estimates of β_i in (7.3).

Table 7.8 shows the estimated propensity scores $\hat{e}(\mathbf{x})$ for four units. $\mathbf{y}$ is a binary vector coded as 2 or 3 if the unit lives in the city or suburbs, respectively, and $\mathbf{x_1}, \ldots, \mathbf{x_7}$ are the seven covariate vectors. $\hat{e}(\mathbf{x})$ is the probability of living in the suburbs. The first unit in the table has a 0.4 probability of living in the suburbs or 0.6 probability of living in the city, and the fourth unit a 0.6 probability of living in the suburbs. Notice that the first and second units live in the city and the third and fourth units in the suburbs respectively, so if we used, say, 0.5 as a cut off then all four units would be classified correctly.

7.6.1 The Balancing Property of the Propensity Score

The balancing property of the propensity score says that treated $(T = 1)$ and control $(T = 0)$ subjects with the same propensity score $e(\mathbf{x})$ have the same distribution of the observed covariates, $\mathbf{x}$,

$$P\left(\mathbf{x}|T = 1, e(\mathbf{x})\right) = P\left(\mathbf{x}|T = 0, e(\mathbf{x})\right)$$

or

$$T \perp \mathbf{x} | e(\mathbf{x}).$$

This means that treatment is independent of the observed covariates conditional on the propensity score.

The balancing property says that if two units, i and j, one of whom is treated, are paired, so that they have the same value of the propensity score $e(\mathbf{x}_i) = e(\mathbf{x}_j)$, then they may have different values of the observed covariate, $\mathbf{x}_i \neq \mathbf{x}_j$, but in this pair the specific value of the observed covariate will be unrelated to the treatment assignment. If many pairs are formed this way, then the the distribution of the observed covariates will look about the same in the treated and control groups, even though individuals in matched pairs will typically have different values of x. This leads to method that matches treated to control subjects based on the propensity score. Matching on $e(\mathbf{x})$ will tend to balance many covariates.

7.6.2   Propensity Score and Ignorable Treatment Assignment

Definition 7.7 (Strongly Ignorable Treatment Assignment). A treatment assignment T is strongly ignorable, if

$$P(T | Y(0), Y(1), \mathbf{x}) = P(T | \mathbf{x}),$$

or

$$T \perp Y(0), Y(1) | \mathbf{x}.$$

It may be difficult to find a treated and a control unit that are closely matched for every one of the many covariates in x, but it is easy to match on one variable, the propensity score, $e(\mathbf{x})$, and doing that will create treated and control groups that have similar distributions for all the covariates.

Ignorable treatment assignment and the balancing property of the propensity score implies that (for a proof see Rosenbaum [2010])

$$P(T | Y(0), Y(1), e(\mathbf{x})) = P(T | e(\mathbf{x})),$$

or

$$T \perp Y(0), Y(1) | e(\mathbf{x}).$$

This means that the scalar propensity score $e(\mathbf{x})$ may be used in place of the many covariates in $\mathbf{x}$.

If treatment assignment is ignorable, then propensity score methods will produce unbiased results of the treatment effects. In Example 7.9 what does it mean for treatment assignment to be ignorable? The potential outcomes for weight gain in the suburbs (treated) and city (control) groups are independent of treatment assignment conditional on the propensity score.

Suppose a critic came along and claimed that the study did not measure an important covariate (e.g., parents' weight) so the study is in no position to claim that the suburban and city groups are comparable. This criticism could be dismissed in a randomized experiment — randomization tends to balance unobserved covariates — but the criticism cannot be dismissed in an observational study. This difference in the unobserved covariate, the critic continues, is the real reason outcomes differ in the treated and control groups: it is not an effect caused by the treatment, but rather a failure on the part of the investigators to measure and control imbalances in the unobserved covariate. The sensitivity of an observational study to bias from an unmeasured covariate is the magnitude of the departure from the model that would need to be present to materially alter the study's conclusions. Statistical methods to measure how sensitive an observational study is to this type of bias are discussed in Rosenbaum [2010] and Imbens and Rubin [2015].

7.7 Propensity Score Methods to Reduce Bias in Observational Studies

If experimental units are randomized to different treatments, then there should be no selection bias (or systematic differences) in observed or unobserved covariates between the treatment groups. In a study where the investigator does not have control over the treatment assignment. a direct comparison could be misleading. However, covariate information is incorporated into the study design or into adjustment of the treatment effect, then a direct comparison might be appropriate. Most standard methods such as stratification and covariance adjustment can only use a limited number of covariates, but propensity scores are a scalar summary of this information and hence don't have this limitation [d'Agostino, 1998].

The primary use of propensity scores in observational studies is to reduce bias and increase precision. The three most common techniques that use the propensity score are matching, stratification (also called subclassification) and regression adjustment. Each of these techniques is a way to make an adjustment for covariates prior to (matching and stratification) or while (stratification and regression adjustment) calculating the treatment effect. With all three techniques, the propensity score is calculated the same way, but once it is estimated, it is applied differently. Propensity scores are useful for these techniques because by definition the propensity score is the conditional probability of treatment given the observed covariates $e(\mathbf{x}) = P(T = 1|X)$, which implies that T and $\mathbf{x}$ are conditionally independent given $e(\mathbf{x})$. Thus, subjects in treatment and control groups with equal (or nearly equal) propensity scores will tend to have the same (or nearly the same) distributions on their background covariates. Exact adjustments made using the propensity score will, on average, remove all of the bias in the background covariates. Therefore, bias-removing adjustments can be made using the propensity scores rather than all of the background covariates individually [d'Agostino, 1998].

7.7.1 Propensity Score Matching

Matching can refer to any method that balances the distribution of covariates in the treated and control groups. For example, we might match on income, and state that a treated unit is close to a control unit if the distance between the treated unit's income and control unit's income is the smallest among all possible control units, although there are other ways to define closeness. If this was done in an observational study comparing a treatment, then the income covariate would be balanced, but other observed covariates may not be balanced. The balancing property of the propensity score suggests that using this scalar score as a way to compare treated and control units and measure distance between two units should be similar to balancing all the covariates that are used to construct the propensity score.

Implementing matching methods has four key steps [Stuart, 2010]:

1. Defining "closeness": the distance used to determine if an individual is a good match for another.

2. Implementing a matching method.

3. Assessing the quality of the matched sample. This may include repeating steps 1 and 2 until a well-matched sample is obtained.

4. Analysis of the outcome and estimation of the treatment effect, given the matching done in step 3.

This section will focus on 1:1 nearest neighbour matching using distance measures based on the propensity score.

7.7.2 Distance Between Units

Let D_{ij} define a measure of similarity between two units units i and j for matching. Two measures based on the propensity score $e_i = e(\mathbf{x_i})$ use either the propensity score difference:

$$D_{ij} = |e_i - e_j|,$$

or the linear propensity score

$$D_{ij} = |\text{logit}(e_i) - \text{logit}(e_j)|,$$

where, $\text{logit}(x) = \log(x/(1-x))$, $x \in (0,1)$.

7.7.3 Nearest Neighbour Matching

Consider the data shown in Table 7.9 with four treated and five control units and propensity scores e_i. A greedy 1:1 nearest neighbour matching algorithm starts with the treated unit i and selects the closest control unit. Starting with unit A, the closest match is unit b, so the first pair is {Ab}, and unit b is removed as a possible control match for the remaining matches; unit B has closest match d, so the second pair is {Bd}, and unit d is removed as a possible control match for the remaining matches; unit C has closest match e, so the third pair is {Ce}, and unit e is removed from the list of possible control matches; finally unit D is matched with c, since c is the closest remaining match.

This matching method is greedy in the sense that each pairing occurs without reference to how other units will be or have been paired, instead of, say, optimizing a specific criterion, and it is 1:1 since each treated subject is matched to one control.

TABLE 7.9: Hypothetical Data to Illustrate Matching

Treated		Control	
Unit	e_i	**Unit**	e_i
A	0.42	a	0.44
B	0.35	b	0.42
C	0.24	c	0.37
D	0.22	d	0.34
		e	0.23

7.7.4 Computation Lab: Nearest Neighbour Propensity Score Matching

The data from Example 7.9 is in the data frame nhefs9282. In this section, the Nearest Neighbour Matching algorithm to match on the propensity score is implemented.

As a first step let's define the data set for analysis that removes any rows with missing values of the covariates,

```
nhefs9282_propmoddat <-
  nhefs9282 %>%
  filter(urbanrural1982 == 2 | urbanrural1982 == 3) %>%
  mutate(id = HANESEQ,
         urbanrural1982 =
             recode(urbanrural1982, `2` = 0, `3` = 1)) %>%
  select(
    id,
    wtgain,
    wt1982,
    wt1992,
    urbanrural1982,
    age1982,
    sex,
    smoke1982,
    genhealth1982,
    physicalactive1982,
    marital1982,
    incomeclass
  ) %>%
  na.omit()
```

and then fit a logistic regression model to estimate the propensity scores. `predict(propmod_nhefs, type = "response")` computes the propensity scores; and `predict(propmod_nhefs, type = "link")` computes the linear propensity scores.

```
propmod_nhefs <-
  glm(
    as.factor(urbanrural1982) ~
      age1982 + as.factor(sex) + as.factor(smoke1982) +
      as.factor(genhealth1982) + as.factor(physicalactive1982) +
      as.factor(marital1982) + as.factor(incomeclass),
    data = nhefs9282_propmoddat,
    family = binomial
  )

psscore <- as.data.frame(predict(propmod_nhefs,
                               type = "response"))
names(psscore) <- "psscore"
```

The propensity scores are then combined with the original data frame using
cbind().

```
nhefs_ps <- cbind(psscore, nhefs9282_propmoddat)
```

Next, the treated (urbanrural1982 == 1) and control (urbanrural1982 ==
0) observations are stored in separate data frames df_trt and df_cnt. The
data in the treatment group is put in descending order from the largest
propensity score. Other orderings are possible, including aescending or random
ordering, but these will yield different matchings.

```
df_trt <-
  nhefs_ps %>%
  filter(urbanrural1982 == 1) %>%
  arrange(desc(psscore)) %>%
  select(id, wtgain, psscore, urbanrural1982)

df_cnt <-
  nhefs_ps %>%
  filter(urbanrural1982 == 0) %>%
  select(id, wtgain, psscore, urbanrural1982)
```

The next code chunk implements the greedy 1:1 nearest neighbour matching
algorithm. pairs_out will store the results of the algorithm.

```
pairs_out <- data.frame(
  id = integer(nrow(df_trt) * 2),
  trt = integer(nrow(df_trt) * 2),
  pair = integer(nrow(df_trt) * 2)
)
```

1. Define the control group. This will change after each iteration, since control
   subjects are removed as they are matched.
2. (a) In the first iteration calculate the $k_1 = \arg\min_j\{|e_{1T} - e_{jC}|, j = 1, \ldots, n_C\}$, where e_{jC} is the jth propensity score in the control group.

(b) Remove control unit with index k_1 from the set of possible control units and calculate $k_2 = \arg \min_j \{|e_{1T} - e_{jC}|, j = 1, \ldots, k_1 - 1, k_1 + 1, \ldots, (n_C - 1)\}$. Proceed in this manner until all the treated units are matched.

3. Save treated-control pairs $\{1, k_1\}, \{2, k_2\}, \ldots$, in a data frame with the outcome variable and information about pairs.

```
for (i in 1:nrow(df_trt)) {
  # 1. define control group

  if (i == 1) {
    df_cnt1 <- df_cnt
  } else if (i > 1) {
    df_cnt1 <- df_cnt1[-k1,]
  }

  # 2. calculate difference between ith ps score in treat
  # and controls that haven't already been used then
  # find index of control with smallest ps score diff

  diff <- abs(df_trt$psscore[i] - df_cnt1$psscore)
  k1 <- which.min(diff)

  # 3. save results in data frame

  # Treatment id
  pairs_out$id[2 * i - 1] <- df_trt$id[i]
  # Control id
  pairs_out$id[2 * i] <- df_cnt1$id[k1]
  # Pair, treatment data
  pairs_out$pair[c(2 * i - 1, 2 * i)] <- i
  pairs_out$trt[c(2 * i - 1, 2 * i)] <- c(1, 0)
}
```

7.7.5 Assessing Covariate Balance in the Matched Sample

How can the degree of balance in the covariate distributions between treated and control units be assessed? The standardized difference between groups in average covariate values by treatment status, scaled by their sample standard deviation provides a scale-free way to assess the balance of each covariate.

Let x be a continuous covariate. $\bar{x}_t$ and S_t^2 are the mean and variance of a covariate in the treated group, and $\bar{x}_c$ and S_c^2 are the mean and variance of a covariate in the control group. The pooled variance is

$$\hat{\sigma}^2 = \sqrt{\frac{S_t^2 + S_c^2}{2}}.$$

If x is a binary variable then $\bar{x}$ is the sample proportion and the pooled variance is

$$\hat{\sigma}^2 = \sqrt{\frac{\hat{p}_t(1 - \hat{p}_t)/n_t + \hat{p}_c(1 - \hat{p}_c)/n_c}{2}}.$$

$\hat{\sigma}^2$ is computed before matching. There are other statistics and graphical methods that can be used to assess covariate balance.

The standardized difference between two groups before matching is

$$\frac{100 \times |\bar{x}_t - \bar{x}_c|}{\hat{\sigma}^2}. \tag{7.4}$$

After matching we replace $\bar{x}_c$ with $\bar{x}$ of matched controls, $\bar{x}_{cm}$ in (7.4). Then it's possible to assess the change in standardized difference before and after matching.

7.7.5.1   Imbalance versus Overlap

Imbalance occurs if the distributions of relevant pre-treatment variables differ for the treatment and control groups.

Lack of complete overlap occurs if there are values of pre-treatment variables where there are treated units but no controls, or controls but no treated units. Lack of complete overlap creates problems because it means that there are treatment observations for which we have no counterfactuals (that is, control observations with the same covariate distribution) and vice versa. When treatment and control groups do not completely overlap, the data are inherently limited in what they can tell us about treatment effects in the regions of nonoverlap. No amount of adjustment can create direct treatment/control comparisons, and one must either restrict inferences to the region of overlap or rely on a model to extrapolate outside this region [Gelman and Hill, 2006].

Propensity score matching discards units with propensity score values outside the range of the other group, so that there is substantial overlap in the propensity score distributions between the two groups.

7.7.6 Computation Lab: Checking Overlap and Covariate Balance

The code chunk below produces Figure 7.1. `nhefs_ps` is joined with `pairs_out`, and then the matched (`!is.na(pair)`) and unmatched units (`is.na(pair)`) are categorized based on treatment (`urbanrural1982`).

```
nhefs_ps %>%
  left_join(pairs_out, by = "id") %>%
  mutate(
    mcc = case_when(
      !is.na(pair) &
        urbanrural1982 == 0 ~ "Matched Control Units",!is.na(pair) &
        urbanrural1982 == 1 ~ "Matched Treatment Units",
      is.na(pair) &
        urbanrural1982 == 0 ~ "Unmatched Control Units",
      is.na(pair) &
        urbanrural1982 == 1 ~ "Unmatched Treatment Units"
    )
  ) %>%
  ggplot(aes(mcc, psscore)) +
  geom_boxplot() +
  labs(x = "Matched/Unmatched Units", y = "Propensity Score")
```

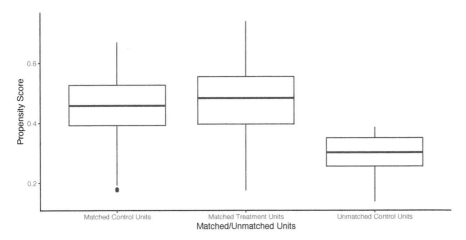

FIGURE 7.1: Distribution of Propensity Scores Using 1:1 Nearest Neighbour Matching on Propensity Score

Figure 7.1 shows the distribution of propensity scores in the matched and unmatched treatment and control groups. There is adequate overlap of the propensity scores with a good control match for each treated unit.

Checking covariate balance before and after the propensity score matching that was conducted in Section 7.7.4 is an important part of the design stage of an observational study that uses propensity score matching. Let's write a function to compute (7.4) before and after matching.

The code chunk below computes the sample means and variances for each treatment group using `group_by(urbanrural1982)` with `summarise(m = mean(age1982), v = var(age1982))`; `ungroup()` then removes the grouping variable; the data frame is widened so that there is a column for each sample mean and variance, and (7.4) is calculated.

```
nhefs9282_propmoddat %>%
  group_by(urbanrural1982) %>%
  summarise(m = mean(age1982),
            v = var(age1982)) %>%
  ungroup() %>%
  pivot_wider(names_from = urbanrural1982,
              values_from = c(m, v)) %>%
  mutate(std_diff = 100 * abs((m_0 - m_1) /
                               sqrt((v_0 + v_1) / 2)))
```

Since we will want to do this with many variables, a function should be written. When programming with data - variables using the `dplyr` library, the argument needs to be embraced by surrounding it in double braces like `{{ var }}`.[1] `checkbal_continuous` turns the code above into a function that can be used to check the covariate balance for a continuous variable in a data frame.

```
checkbal_continuous <- function(data, var1, trt) {
  data %>%
    group_by({{ trt }}) %>%
    summarise(m = mean({{ var1 }}),
              v = var({{ var1 }})) %>%
    ungroup() %>%
    pivot_wider(names_from = {{ trt }},
                values_from = c(m, v)) %>%
    mutate(std_diff = 100 * abs((m_0 - m_1) /
                                 sqrt((v_0 + v_1) / 2)))
}
```

---

[1] https://dplyr.tidyverse.org/articles/programming.html#fn2

We can use it to check the balance of age1982 before and after matching.

```
# before matching
checkbal_continuous(data = nhefs9282_propmoddat,
                    var1 = age1982,
                    trt = urbanrural1982)
```

```
## # A tibble: 1 x 5
##     m_0   m_1   v_0   v_1 std_diff
##   <dbl> <dbl> <dbl> <dbl>    <dbl>
## 1  52.1  49.5  137.  119.     22.7
```

```
# after matching
nhefs9282_propmoddat %>%
  left_join(pairs_out, by = "id") %>%
  filter(!is.na(pair)) %>%
  checkbal_continuous(age1982, urbanrural1982)
```

```
## # A tibble: 1 x 5
##     m_0   m_1   v_0   v_1 std_diff
##   <dbl> <dbl> <dbl> <dbl>    <dbl>
## 1  50.6  49.5  118.  119.     9.67
```

In this case, the percent improvement in balance is $100 \times (22.67 - 9.67)/22.67 = 57.3445$. Values greater than 0 indicate an improvement and values less than 0 indicate that the balance is worse after matching.

checkbal_discrete() is a function that can be used to check balance for discrete covariates.

```
checkbal_discrete <- function(data, trt, var){
  data %>%
    group_by({{ trt }}, {{ var }}) %>%
    count({{ var }}) %>%
    ungroup({{ var }}) %>%
    mutate(t = sum(n),
           p = n/t,
           v = p*(1-p)/t) %>%
    select({{ var }}, {{ trt }}, p, v) %>%
    pivot_wider(names_from = {{ trt }},
                values_from = c(p,v)) %>%
    mutate(std_diff = 100 *
           abs((p_1 - p_0) / sqrt((v_0 + v_1) / 2)))
}
```

```
# before matching
nhefs9282_propmoddat %>% checkbal_discrete(urbanrural1982, sex)
```

```
## # A tibble: 2 x 6
##     sex   p_0   p_1      v_0       v_1 std_diff
##   <dbl> <dbl> <dbl>    <dbl>     <dbl>    <dbl>
## 1     1 0.433 0.432 0.000178 0.000224     1.08
## 2     2 0.567 0.568 0.000178 0.000224     1.08
```

```
# after matching
nhefs9282_propmoddat %>%
  left_join(pairs_out, by = "id") %>%
  filter(!is.na(pair)) %>%
  checkbal_discrete(urbanrural1982, sex)
```

```
## # A tibble: 2 x 6
##     sex   p_0   p_1      v_0       v_1 std_diff
##   <dbl> <dbl> <dbl>    <dbl>     <dbl>    <dbl>
## 1     1 0.433 0.432 0.000224 0.000224     6.10
## 2     2 0.567 0.568 0.000224 0.000224     6.10
```

The balance for sex decreased after matching since $100 \times (1.08 - 6.10)/1.08 =$ -464.8148. The large value should not be too concerning since the standardized difference changed from 1 to 6 which are both fairly low.

7.7.7   Analysis of the Outcome after Matching

The analysis should proceed as if the treated and control samples had been generated through randomization. Individual units will be well matched on the propensity score, but may not be matched on the full set of covariates. Thus, it is common to pool all the matches into matched treated and control groups and run analyses using these groups rather than individual matched pairs [Stuart, 2010].

Consider Example 7.9. Let y_i^{1982} and y_i^{1992} be the ith subjects' weights at baseline in 1982 and 1992. Two common statistical models to evaluate the effect of treatment T_i on weight gain are:

$$y_i^d = \beta_0 + \beta_1 T_i + \epsilon_i \tag{7.5}$$

$$y_i^{1992} = \beta_0 + \beta_1 T_i + \beta_2 y_i^{1982} + \epsilon_i, \tag{7.6}$$

where $y_i^d = y_i^{1992} - y_i^{1982}$, $\epsilon_i \sim N(0, \sigma^2)$, and T_i is the treatment indicator for the ith unit. Propensity score matching allows us to define T_i as an indicator of treated $T = 1$ or control $T = 0$ based on propensity score matching. If $\beta_2 = 1$ in (7.6) then models (7.6) and (7.5) are equivalent.

7.7.8   Computation Lab: Analysis of Outcome after Matching

Analysis of the outcome is the final stage of implementing propensity score matching. Continuing with Example 7.9, by using the propensity score matching from Section 7.7.4 we can compare weight gain in people living in the city versus the suburbs.

The treatment and control groups from propensity score matching were created in Section 7.7.4 and stored in the data frame `pairs_out`. `nhefs_ps` is a data frame that contains the outcome variables, covariates, and propensity scores. It can be merged with the data on matching to create a data frame, `df_anal`, ready for analysis.

```
df_anal <- nhefs_ps %>%
  left_join(pairs_out, by = "id") %>%
  mutate(
    treat =
      case_when(
        !is.na(pair) & urbanrural1982 == 0 ~
          "Matched Control Units",
        !is.na(pair) & urbanrural1982 == 1 ~
          "Matched Treatment Units"
      )
  )
```

Fitting model (7.5),

```
mod_ttest_ps <- lm(wtgain ~ treat, data = df_anal)
broom::tidy(mod_ttest_ps, conf.int = TRUE)
```

```
## # A tibble: 2 x 7
##   term     estimate std.error statistic p.value conf.low
##   <chr>       <dbl>     <dbl>     <dbl>   <dbl>    <dbl>
## 1 (Inter~      3.12     0.523      5.97 2.69e-9     2.10
## 2 treatM~     0.801     0.739      1.08 2.79e-1   -0.649
## # ... with 1 more variable: conf.high <dbl>
```

The estimate for the treatment effect is $\bar{y}_{Treat}^{d} - \bar{y}_{Control}^{d} = 0.8$. The estimate can also be calculated directly from the treatment means, but using lm() also computes inferential statistics.

```
df_anal %>%
  group_by(treat) %>%
  summarise(ybar_d = mean(wtgain),
            ybar_1982 = mean(wt1982),
            ybar_1992 = mean(wt1992))
```

```
## # A tibble: 3 x 4
##   treat                   ybar_d ybar_1982 ybar_1992
##   <chr>                    <dbl>     <dbl>     <dbl>
## 1 Matched Control Units     3.12      165.      168.
## 2 Matched Treatment Units   3.92      161.      165.
## 3 <NA>                     0.842      161.      162.
```

Fitting model (7.6),

```
mod_ancova <- lm(wt1992 ~ wt1982 + treat, data = df_anal)
broom::tidy(mod_ancova)
```

```
## # A tibble: 3 x 5
##   term            estimate std.error statistic  p.value
##   <chr>              <dbl>     <dbl>     <dbl>    <dbl>
## 1 (Intercept)         16.7      1.86      8.97 6.21e-19
## 2 wt1982             0.918    0.0108      84.7 0
## 3 treatMatched T~    0.512     0.731     0.700 4.84e- 1
```

The treatment effect in this model is the difference between treated and control adjusted mean weights.

How does this compare to not using propensity score matching? We can fit models (7.5) and (7.6) using **urbanrural1982** as the treatment indicator.

```
mod_ttest <- lm(wtgain ~ urbanrural1982, data = df_anal)
broom::tidy(mod_ttest)
```

```
## # A tibble: 2 x 5
##    term            estimate std.error statistic   p.value
##    <chr>              <dbl>     <dbl>     <dbl>     <dbl>
## 1 (Intercept)         2.65     0.480      5.53   3.52e-8
## 2 urbanrural1982      1.27     0.721      1.76   7.81e-2
```

```
mod_ttest <- lm(wt1992 ~ wt1982 + urbanrural1982, data = df_anal)
broom::tidy(mod_ttest)
```

```
## # A tibble: 3 x 5
##    term            estimate std.error statistic   p.value
##    <chr>              <dbl>     <dbl>     <dbl>     <dbl>
## 1 (Intercept)        16.8      1.78       9.45 7.82e-21
## 2 wt1982              0.914    0.0105     87.3 0
## 3 urbanrural1982      1.03     0.712      1.45 1.48e- 1
```

The treatment effects are larger in the unadjusted analyses, but neither analyses indicate that there is evidence of a meaningful or statistically significant effect related to living in the suburbs versus the city.

7.7.9   Subclassification

7.7.9.1   Subclassification as a Method to Remove Bias

Consider a researcher who wishes to compare a response variable y in two treatment groups, but is concerned that the differences in the mean values of y in the groups may be due to other covariates $x_1, x_2, \ldots$ in which the groups differ, rather than treatment T. Cochran [1968] shows that if the distribution of a single covariate x, that is independent of T but associated with y, is broken up into five or more subclasses (e.g., using the quintiles) then 90% of the bias due to x is removed.

Example 7.11 (Smoking and Lung Cancer). The following data were selected from data supplied to the U. S. Surgeon General's Committee from three of the studies in which comparisons of the death rates of men with different smoking habits were made [Cochran, 1968]. Table 7.10 shows the unadjusted death rates per 1,000 person-years. We might conclude that cigar and pipe smokers should give up smoking, but if they lack the strength of will to do so, they should switch to cigarettes.

TABLE 7.10: Unadjusted Death Rates per 1,000 Person-years

Smoking group	Canadian	British	U.S.
Non-smokers	20.2	11.3	13.5
Cigarettes only	20.5	14.1	13.5
Cigars, pipes	35.5	20.7	17.4

Are there other variables in which the three groups of smokers may differ, that are related to the probability of dying; and that are not themselves affected by smoking habits?

The regression of probability of dying on age for men over 40 is a concave upwards curve, the slope rising more and more steeply as age advances. The mean ages for each group in the previous table are as follows. Table 7.11 shows that cigar or pipe smokers are older on average than the men in the other groups. In both the U.S. and Canadian studies which showed similar death rates in smokers and non-smokers, the cigarette smokers are youger than the non-smokers.

TABLE 7.11: Mean Age (in Years) of Men in Each Group

Smoking group	Canadian	British	U.S.
Non-smokers	54.9	49.1	57.0
Cigarettes only	50.5	49.8	53.2
Cigars, pipes	65.9	55.7	59.7

Table 7.12 shows the adjusted death rates obtained when the age distributions were divided into 9-11 subclasses (the maximum number of subclasses allowed by the data). The adjusted cigarette death rates now show a substantial increase over non-smoker death rates in all the studies.

TABLE 7.12: Adjusted Death Rates Using 9-11 Subclasses

Smoking group	Canadian	British	U.S.
Non-smokers	20.2	11.3	13.5
Cigarettes only	29.5	14.8	21.2
Cigars, pipes	19.8	11.0	13.7

7.7.9.2 Subclassification and the Propensity Score

Rosenbaum and Rubin [1984] extended the idea of subclassification to the propensity score and showed that creating subclasses based on the quintiles of the propensity score eliminated approximately 90% of the bias due to covariates used in developing the propensity score. One advantage of this method, compared to say matching on the propensity score, is that the whole sample is used and not just matched sets.

Covariate balance can be checked by comparing the distribution of covariates before and after subclassification. We can compare the treated and control groups on the covariates x_i used to build the propensity score model before and after stratification. The distributions before stratification can be evaluated by fitting a linear model.

$$\mu_i = E(x_i) = \beta_0 + \beta_1 T_i. \tag{7.7}$$

The distributions after stratification can be evaluated by fitting

$$\mu_i = E(x_i) = \beta_0 + \beta_1 T_i + \beta_2 Q_i + \beta_3 T_i Q_i, \tag{7.8}$$

where Q_i is the propensity score quintile for the ith unit. The distributions for each covariate can be compared using the square of the t-statistic (or F statistic) for testing main effects of T in (7.7) and (7.8).

Once adequate covariate balance has been achieved using propensity score subclassification, the analysis of the main outcome y can be evaluated by separately computing the treatment effect within each propensity score quintile. The propensity score quintile can also be added as a covariate in order to compute an overall treatment effect adjusted for propensity score subclass.

7.7.10 Computation Lab: Subclassification and the Propensity Score

This section will illustrate subclassification using the propensity score with data from Example 7.9. The propensity scores and binary indicator variables for categorical covariates can be extracted from the propensity score model object `propmod_nhefs`.

The quintiles of the propensity score can be found using `quantile()`.

```
ps_strat <-
  quantile(propmod_nhefs$fitted.values, c(0.2, 0.4, 0.6, 0.8))
ps_strat
```

```
##     20%    40%    60%    80%
## 0.3423 0.4169 0.4825 0.5482
```

The first subclass consists of units whose propensity score are less than or equal to 0.3423, the second subclass consists of units whose propensity score is greater than 0.3423 but less than or equal to 0.4169, etc. We can create the subclasses in R using `mutate()`. But, first we will extract the `model.matrix` from the propensity score model `propmod_nhefs` so that the covariates used to build the propensity score model are the dummy variables coded as 0/1 for reasons that will be explained below.

Covariates such as `sex` have been coded as as.factor(sex)2—`as.factor(sex)` coerces `sex` from a numeric variable to a factor variable using the `contr.treatment()` contrast function, where the number appended to the end of the column name is the baseline level, which in this case is female, since `sex == 2` is female.

`nhefs_psstratdf` includes the covariates used in the model and propensity scores.

```
nhefs_psstratdf <-
  data.frame(
    cbind(
      propmod_nhefs$data$id,
      propmod_nhefs$y,
      propmod_nhefs$fitted.values,
      model.matrix(propmod_nhefs)[, 2:18]
    )
  )

colnames(nhefs_psstratdf) <-
  c("id", "urbanrural1982", "psscore",
    colnames(model.matrix(propmod_nhefs))[2:18])
```

`nhefs_psstratdf` is used to define propensity score strata based on `ps_strat`.

```
nhefs_ps <- nhefs_psstratdf %>%
  mutate(ps_subclass = case_when((psscore <= ps_strat[1]) ~ 1,
                                 (psscore > ps_strat[1] &
                                     psscore <= ps_strat[2]) ~ 2,
                                 (psscore > ps_strat[2] &
                                     psscore <= ps_strat[3]) ~ 3,
                                 (psscore > ps_strat[3] &
                                     psscore <= ps_strat[4]) ~ 4,
                                 (psscore > ps_strat[4]) ~ 5
 ))
```

The distribution of treated and control subjects within each subclass is obtained by grouping treatment and control using `group_by()`, counting using `count()`, and then applying `pivot_wider()` to the data frame to make the table easier to read.

```
nhefs_ps %>%
  group_by(ps_subclass, urbanrural1982) %>%
  count() %>%
  pivot_wider(names_from = ps_subclass, values_from = n)
```

```
## # A tibble: 2 x 6
## # Groups:   urbanrural1982 [2]
##   urbanrural1982   `1`   `2`   `3`   `4`   `5`
##            <dbl> <int> <int> <int> <int> <int>
## 1              0   344   310   283   240   201
## 2              1   151   184   211   255   293
```

Since 6 of the 7 covariates are categorical variables, they are represented as $k - 1$ dummy variables, where k is the number of categories, in the propensity score model, resulting in a total of 17 covariates that will be compared.

An efficient way to code the computation for fitting the models and extract the F statistics, without having to write out 17 model statements twice, is to use `purrr::map()` or `lapply()`.

```
bal_prestrat <- nhefs_ps %>%
  select(4:20) %>%
  map(~ glm(.x ~ nhefs_ps$urbanrural1982, data = nhefs_ps)) %>%
  map(summary) %>%
  map(coefficients)
```

nhefs_ps %>% select(4:20) selects the covariates from nhefs_ps according to their column position and then the models are fit using map(~ glm(.x ~ nhefs_ps$urbanrural1982, data = nhefs_ps)) (.x is the argument for the columns selected in nhefs_ps), computes the summaries of each model map(summary) (which contain the desired t-statistic for T), and finally extracts the coefficients map(coefficients).

```
Fprestrat <- map(1:17, ~ bal_prestrat[[.x]][2, 3]) %>%
  map( ~ round(.x ^ 2, 4)) %>%
  flatten_dbl()
```

The t-statistics are in the second row, third column of each model summary in the list of 17 model summaries bal_prestrat. map(1:17, ~bal_prestrat[[.x]][2,3]) extracts the t-statistics into a list, squares and rounds each t-statistic to compute the F-statistic, and then converts the list to a vector purrr::flatten_dbl().

```
bal_poststrat <- nhefs_ps %>%
  select(4:20) %>%
  map( ~ glm(.x ~ nhefs_ps$urbanrural1982 * nhefs_ps$ps_subclass,
            data = nhefs_ps)) %>%
  map(summary) %>%
  map(coefficients)

Fpoststrat <- map(1:17, ~ bal_poststrat[[.x]][2, 3]) %>%
  map(~ round(.x ^ 2, 4)) %>%
  flatten_dbl()
```

We can examine the distribution of F statistics before stratification,

```
summary(Fprestrat)
```

```
##    Min. 1st Qu.  Median    Mean 3rd Qu.    Max.
##    0.00    1.98    6.62    9.70   12.44   44.94
```

and after stratification.

```
summary(Fpoststrat)
```

```
##     Min. 1st Qu.  Median    Mean 3rd Qu.     Max.
##    0.011   0.287   0.994   2.620   1.725   20.578
```

The median of the F statistics decreased from 6.62 to 0.994 suggesting that stratification has increased covariate balance, although not all the covariates' balance improved.

The treatment effects for each subclass are computed below.

```
psstrat_df <- nhefs_ps %>%
  select(id, ps_subclass) %>%
  left_join(nhefs9282_propmoddat, by = "id") %>%
  select(ps_subclass, wtgain, wt1982, wt1992, urbanrural1982)

psstrat_df %>%
  split(psstrat_df$ps_subclass) %>%
  map( ~ lm(wtgain ~ urbanrural1982, data = .)) %>%
  map(summary) %>% map(coefficients)
```

```
## $`1`
##                 Estimate Std. Error t value Pr(>|t|)
## (Intercept)      -1.6483      1.089 -1.5133   0.1309
## urbanrural1982   -0.6696      1.972 -0.3396   0.7343
##
## $`2`
##                 Estimate Std. Error t value Pr(>|t|)
## (Intercept)        1.477      1.066   1.386   0.1663
## urbanrural1982     1.887      1.746   1.080   0.2805
##
## $`3`
##                 Estimate Std. Error t value Pr(>|t|)
## (Intercept)       3.0919      1.023   3.022 0.002641
## urbanrural1982   -0.6559      1.565  -0.419 0.675422
##
## $`4`
##                 Estimate Std. Error t value  Pr(>|t|)
## (Intercept)       5.7708      1.039   5.555 4.539e-08
## urbanrural1982    0.7468      1.447   0.516 6.061e-01
```

```
##
## $`5`
##                 Estimate Std. Error t value  Pr(>|t|)
## (Intercept)        7.488       1.062  7.0484 6.152e-12
## urbanrural1982    -1.180       1.379 -0.8557 3.926e-01
```

Instead of using wtgain as the outcome, wt1992 and wt1982 could be used in an ANCOVA model.

7.8 Exercises

Exercise 7.1. Suppose N experimental units are randomly assigned to treatment or control by tossing a coin. A unit is assigned to treatment if the coin toss comes up heads. Assume that the probability of tossing a head is p.

a. If $p = 1/3$, then are all treatment assignments equally likely?

b. Identify treatment assignments that do not allow estimating the causal effect of treatment versus control. Explicitly state the treatment assignment vectors for the identified assignments. How many of them are there?

Exercise 7.2. Consider the two treatment assignment mechanisms (7.1) and (7.2). What is the probability $P(T_i = 1)$ in each case? What is $P(T_i = 1|Y_i(0) > Y_i(1))$? Explain how you arrived at your answers.

Exercise 7.3. Consider the data shown in Table 7.4 in Example 7.2. Compute the randomization distribution of the mean difference based on the following treatment assignment mechanisms:

a. Completely randomized where 2 units are assigned to treatments randomly;

b. Randomized based on the mechanism described in Example 7.7 where the random selection depends on the potential outcomes. Does adding randomness to the treatment assignment remove the effect of the confounded treatment assignment in (7.1)?

Exercise 7.4. (Adapted from Gelman and Hill [2006], Chapter 9) Suppose you are asked to design a study to evaluate the effect of the presence of vending machines in schools on childhood obesity. Describe randomized and non-randomized studies to evaluate this question. Which study seems more feasible and easier to implement?

Exercise 7.5. (Adapted from Gelman and Hill [2006], Chapter 9) The table below describes a hypothetical experiment on 2400 persons. Each row of the table specifies a category of person, as defined by their pre-treatment predictor x, treatment indicator T, and potential outcomes $Y(0)$ and $Y(1)$.

Category	Persons in category	x	T	$Y(0)$	$Y(1)$
1	300	0	0	4	6
2	300	1	0	4	6
3	500	0	1	4	6
4	500	1	1	4	6
5	200	0	0	10	12
6	200	1	0	10	12
7	200	0	1	10	12
8	200	1	1	10	12

In making the table we are assuming omniscience, so that we know both $Y(0)$ and $Y(1)$ for all observations. But the (non-omniscient) investigator would only observe x, T, and $Y(T)$ for each unit. For example, a person in category 1 would have $x = 0$, $T = 0$, and $Y(0) = 4$, and a person in category 3 would have $x = 0$, $T = 1$, and $Y(1) = 6$.

a. Give an example of a context for this study. Define $x, T, Y(0), Y(1)$.

b. Calculate the the average causal treatment effect using R.

c. Is the covariate x balanced between the treatment groups?

d. Is it plausible to believe that these data came from a randomized experiment? Defend your answer.

e. Another population quantity is the mean of Y for those who received the treatment minus the mean of Y for those who did not. What is the relation between this quantity and the average causal treatment effect?

f. For these data, is it plausible to believe that treatment assignment is ignorable given the covariate x? Defend your answer.

Exercise 7.6. What is the definition of ignorable treatment assignment?

a. Give an example of a study where the treatment assignment is ignorable.

b. Give an example of a study where the treatment assignment is non-ignorable.

Exercise 7.7. (Adapted from Gelman and Hill [2006], Chapter 9) An observational study to evaluate the effectiveness of supplementing a reading program with a television show was conducted in several schools in grade

4. Some classroom teachers chose to supplement their reading program with the television show and some teachers chose not to supplement their reading program. Some teachers chose to supplement if they felt that it would help their class improve their reading scores. The study collected data on a large number of student and teacher covariates measured before the teachers chose to supplement or not supplement their reading program. The outcome measure of interest is student reading scores.

a. Describe how this study could have been conducted as a randomized experiment.

b. Is it plausible to assume that supplementing the reading program is ignorable in this observational study?

Exercise 7.8. The following questions refer to Section 7.7.4 Computation Lab: Nearest Neighbour Propensity Score Matching.

a. When matching the treatment and control groups, which.min(diff) was used in code chunk below. What does the function which.min() return? Explain the purpose of the line.

```
for (i in 1:nrow(df_trt)) {
  if (i == 1) {
    df_cnt1 <- df_cnt
  } else if (i > 1) {
    df_cnt1 <- df_cnt1[-k1,]
  }

  diff <- abs(df_trt$psscore[i] - df_cnt1$psscore)
  k1 <- which.min(diff)

  pairs_out$id[2 * i - 1] <- df_trt$id[i]
  pairs_out$id[2 * i] <- df_cnt1$id[k1]
  pairs_out$pair[c(2 * i - 1, 2 * i)] <- i
  pairs_out$trt[c(2 * i - 1, 2 * i)] <- c(1, 0)
}
```

b. 2 * i -1 and 2 * i place i^{th} matched treatment and control units respectively. Explain why.

c. Modify the matching code to use the logit of the propensity score, and run the matching algorithm again. Do the results change?

Exercise 7.9. The following questions refer to Example 7.9 and Section 7.7.8.

a. Evaluate the parallelism assumption in `mod_ancova`. What do you conclude?

b. Compute the adjusted difference in mean weight gain using `mod_ancova`. Verify that you obtain the same value as the parameter estimate when using sample means.

Exercise 7.10. Consider the data from Example 7.9. In Section 7.7.4 Computation Lab: Nearest Neighbour Propensity Score Matching,

`marital1982` was included when fitting the propensity score model `propmod_nhefs`. Suppose a (naive) investigator forgot to include `marital1982` in a propensity score model.

a. Fit a new propensity score model without `marital1982`.

b. Use the propensity scores from a. to match treatment and control units using nearest neighbour matching.

c. Check covariate balance post matching and compare with the results from Section 7.7.4 Computation Lab: Checking Overlap and Covariate Balance.

d. Estimate the treatment effect using the model specified in Equation (7.5) and compare with the results from Section 7.7.4 Computation Lab: Analysis of Outcome after Matching. How does including `marital1982` alter the estimate of treatment effect.

Exercise 7.11. In Section 7.7.4 Computation Lab: Nearest Neighbour Propensity Score Matching, a propensity score model for the data from Example 7.10 was fit. The model is stored in `propmod_nhef` and the data is stored in `nhefs9282_propmoddat`.

a. Extract the model matrix from `propmod_nhefs` using `model.matrix()`.

b. Verify that the sex covariate value 1 from `nhefs9282_propmoddat` is mapped to value 0 of the dummy variable in the model matrix and the covariate value 2 is mapped to 1. You can verify the mapping by crosstabulating the original covariate values and the dummy variable values.

Exercise 7.12. Consider the data from Example 7.10 and the propensity score model fitted in Section 7.7.4 Computation Lab: Nearest Neighbour Propensity Score Matching.

a. Create five subclasses based on the quintiles of the propensity score.

b. Within each propensity score subclass, fit the ANCOVA model as described in Equation (7.6). Compare the results you obtain with results from Section 7.7.10 Computation Lab: Subclassification and the Propensity Score.

c. Fit the models to estimate the overall treatment effect adjusting for the propensity score subclasses as shown in Equation (7.9) and Equation (7.10). Interpret your findings.

$$y_i^d = \beta_0 + \beta_1 T_i + \beta_2 S_i \epsilon_i \tag{7.9}$$

$$y_i^{1992} = \beta_0 + \beta_1 T_i + \beta_2 S_i + \beta_3 y_i^{1982} + \epsilon_i, \tag{7.10}$$

where $y_i^d = y_i^{1992} - y_i^{1982}$, $\epsilon_i \sim N(0, \sigma^2)$, T_i is the treatment indicator, and S_i is the propensity score subclass for the ith unit.

d. Estimate the overall treatment effect using the five subclass treatment effects from part b. Compare the estimates with the results from part c.

Bibliography

Adam W Amundson, Rebecca L Johnson, Matthew P Abdel, Carlos B Mantilla, Jason K Panchamia, Michael J Taunton, Michael E Kralovec, James R Hebl, Darrell R Schroeder, Mark W Pagnano, et al. A three-arm randomized clinical trial comparing continuous femoral plus single-injection sciatic peripheral nerve blocks versus periarticular injection with ropivacaine or liposomal bupivacaine for patients undergoing total knee arthroplasty. *Anesthesiology*, 126(6):1139–1150, 2017.

CI Bliss and CL Rose. The assay of parathyroid extract from the serum calcium of dogs. *American Journal of Epidemiology*, 31(3):79–98, 1940.

George EP Box, J Stuart Hunter, and William Gordon Hunter. *Statistics for experimenters: design, innovation, and discovery*, volume 2. Wiley-Interscience New York, 2005.

Alessio Bucciarelli, Gabriele Greco, Ilaria Corridori, Nicola M. Pugno, and Antonella Motta. A design of experiment rational optimization of the degumming process and its impact on the silk fibroin properties. *ACS Biomaterials Science & Engineering*, 7(4):1374–1393, feb 2021. doi: 10.102 1/acsbiomaterials.0c01657. URL https://doi.org/10.1021%2Facsbiomaterials.0c01657.

CDC. Epidemiologic followup study (nhefs), September 2021. URL https://wwwn.cdc.gov/nchs/nhanes/nhefs/default.aspx.

Clinicaltrial.gov. Evaluating the efficacy of hydroxychloroquine and azithromycin to prevent hospitalization or death in persons with covid-19, August 2020. URL https://clinicaltrials.gov/ct2/show/NCT04358068.

William G Cochran. The effectiveness of adjustment by subclassification in removing bias in observational studies. *Biometrics*, pages 295–313, 1968.

Jacob Cohen. A power primer. *Psychological bulletin*, 112(1):155, 1992.

Ralph B d'Agostino. Tutorial in biostatistics: Propensity score methods for bias reduction in the comparison of a treatment to a non-randomized control group. *Stat Med*, 17(19):2265–2281, 1998.

R. Doll and A. B. Hill. Smoking and carcinoma of the lung. *BMJ*, 2(4682): 739–748, sep 1950. doi: 10.1136/bmj.2.4682.739. URL https://doi.org/ 10.1136%2Fbmj.2.4682.739.

Michael D Ernst. Permutation methods: a basis for exact inference. *Statistical Science*, 19(4):676–685, 2004.

Sam Firke. *janitor: Simple tools for examining and cleaning dirty data*, 2021. URL https://github.com/sfirke/janitor. R package version 2.1.0.

Ronald Aylmer Fisher. *The design of experiments*. Number 2nd Ed. Oliver & Boyd, Edinburgh & London., 1937.

Paul H Garthwaite. Confidence intervals from randomization tests. *Biometrics*, pages 1387–1393, 1996.

Andrew Gelman and Jennifer Hill. *Data analysis using regression and multi-level/hierarchical models*. Cambridge University Press, 2006.

Harvard University, Department of Statistics. Xiao-Li Meng is Chicago Statistician of the Year, October 2015. URL http://statistics.fas.harvard .edu/news/xiao-li-meng-chicago-statistician-year.

John F Helliwell, Richard Layard, and Jeffrey Sachs. World happiness report 2019. New York: Sustainable development solutions network. *URL: https://worldhappiness.report/ed/2019/*, 2019.

Paul W Holland. Statistics and causal inference. *Journal of the American Statistical Association*, 81(396):945–960, 1986.

Harold Hotelling. Some improvements in weighing and other experimental techniques. *The Annals of Mathematical Statistics*, 15(3):297–306, 1944.

Guido W Imbens and Donald B Rubin. *Causal inference in statistics, social, and biomedical sciences*. Cambridge University Press, 2015.

Jessica Jaynes, Xianting Ding, Hongquan Xu, Weng Kee Wong, and Chih-Ming Ho. Application of fractional factorial designs to study drug combinations. *Statistics in Medicine*, 32(2):307–318, 2013.

Manasigan Kanchanachitra, Chalermpol Chamchan, Churnrurtai Kanchana-chitra, Kanyapat Suttikasem, Laura Gunn, and Ivo Vlaev. Nudge interventions to reduce fish sauce consumption in Thailand. *PLOS ONE*, 15 (9):1–18, 09 2020. doi: 10.1371/journal.pone.0238642. URL https://doi.org/10.1371/journal.pone.0238642.

Young-Woo Kim, Jae-Moon Bae, Young-Kyu Park, Han-Kwang Yang, Wansik Yu, Jeong Hwan Yook, Sung Hoon Noh, Mira Han, Keun Won Ryu, Tae Sung Sohn, Hyuk-Joon Lee, Oh Kyoung Kwon, Seung Yeob Ryu, Jun-Ho Lee, Sung Kim, Hong Man Yoon, Bang Wool Eom, Min-Gew Choi, Beom Su Kim, Oh Jeong, Yun-Suhk Suh, Moon-Won Yoo, In Seob Lee, Mi Ran Jung,

Ji Yeong An, Hyoung-Il Kim, Youngsook Kim, Hannah Yang, Byung-Ho Nam, and FAIRY Study Group. Effect of intravenous ferric carboxymaltose on hemoglobin response among patients with acute isovolemic anemia following gastrectomy: The fairy randomized clinical trial. *JAMA*, 317(20):2097–2104, May 2017. doi: 10.1001/jama.2017.5703.

Russell V Lenth. Quick and easy analysis of unreplicated factorials. *Technometrics*, 31(4):469–473, 1989.

Russell V. Lenth. *emmeans: Estimated marginal means, aka least-squares means*, 2021. URL https://github.com/rvlenth/emmeans. R package version 1.6.3.

Frederic M Lord. A paradox in the interpretation of group comparisons. *Psychological Bulletin*, 68(5):304, 1967.

Tom M McLellan, John A Caldwell, and Harris R Lieberman. A review of caffeine's effects on cognitive, physical and occupational performance. *Neuroscience & Biobehavioral Reviews*, 71:294–312, 2016.

X.L. Meng. A trio of inference problems that could win you a Nobel Prize in statistics (if you help fund it). In Xihong Lin, Christian Genest, David L Banks, Geert Molenberghs, David W Scott, and Jane-Ling Wang, editors, *Past, present, and future of statistical science*, chapter 45, pages 537–557. CRC Press, 2014. URL https://web.archive.org/web/20170815080043 id_/https://dash.harvard.edu/bitstream/handle/1/11718181/COPSS _50.pdf?sequence=1.

Kirill Müller and Hadley Wickham. *tibble: Simple data frames*, 2021. URL https://CRAN.R-project.org/package=tibble. R package version 3.1.4.

Obstetrics and Gynaecology. *Hospital statistics*. University of Toronto, Toronto, Canada, July 2019.

ourworldindata.org. *Life satisfaction vs. child mortality*, 2021. URL https://ourworldindata.org/happiness-and-life-satisfaction.

Thomas Lin Pedersen. *patchwork: The composer of plots*, 2020. URL https://CRAN.R-project.org/package=patchwork. R package version 1.1.1.

R Core Team. *R: A language and environment for statistical computing*. R Foundation for Statistical Computing, Vienna, Austria, 2021. URL https://www.R-project.org/.

Paul R Rosenbaum. Design of observational studies. *New York, USA: Springer. doi*, 10:978–1, 2010.

Paul R Rosenbaum and Donald B Rubin. Reducing bias in observational studies using subclassification on the propensity score. *Journal of the American Statistical Association*, 79(387):516–524, 1984.

Donald B Rubin. The design versus the analysis of observational studies for causal effects: Parallels with the design of randomized trials. *Statistics in Medicine*, 26(1):20–36, 2007.

Victor GF Santos, Vander RF Santos, Leandro JC Felippe, Jose W Almeida Jr, Rômulo Bertuzzi, Maria APDM Kiss, and Adriano E Lima-Silva. Caffeine reduces reaction time and improves performance in simulated-contest of taekwondo. *Nutrients*, 6(2):637–649, 2014.

Statistics Canada, The Daily. Canada's population estimates, first quarter 2015, June 2015. URL http://www.statcan.gc.ca/daily-quotidien/15 0617/dq150617c-eng.htm.

Elizabeth A. Stuart. Matching methods for causal inference: A review and a look forward. *Statistical Science*, 25(1):1 – 21, 2010. doi: 10.1214/09-STS313. URL https://doi.org/10.1214/09-STS313.

Uthayanath Suthakar, Luca Magnoni, David Ryan Smith, Akram Khan, and Julia Andreeva. An efficient strategy for the collection and storage of large volumes of data for computation. *Journal of Big Data*, 3(1):21, 2016. doi: 10.1186/s40537-016-0056-1. URL https://doi.org/10.1186/s40537-016-0056-1.

Techcrunch. How big is Facebook's data? 2.5 billion pieces of content and 500+ terabytes ingested every day, August 2012. URL https://techcrunch.com /2012/08/22/how-big-is-facebooks-data-2-5-billion-pieces-of-content-and-500-terabytes-ingested-every-day/.

Zeynep Tufekci. Show me the data! *New York Times*, August 2021. URL https://www.nytimes.com/2021/08/27/opinion/covid-data-vaccines.html.

Howard Wainer and Lisa M Brown. Two statistical paradoxes in the interpretation of group differences: Illustrated with medical school admission and licensing data. *The American Statistician*, 58(2):117–123, 2004.

Hadley Wickham. *Advanced R*. CRC press, 2019.

Hadley Wickham. *tidyverse: Easily install and load the tidyverse*, 2021. URL https://CRAN.R-project.org/package=tidyverse. R package version 1.3.1.

Hadley Wickham and Garrett Grolemund. *R for data science: import, tidy, transform, visualize, and model data*. O'Reilly Media, Inc., 2016.

Hadley Wickham, Winston Chang, Lionel Henry, Thomas Lin Pedersen, Kohske Takahashi, Claus Wilke, Kara Woo, Hiroaki Yutani, and Dewey Dunnington. *ggplot2: Create elegant data visualisations using the grammar of graphics*, 2021a. URL https://CRAN.R-project.org/package=ggplot2. R package version 3.3.5.

Hadley Wickham, Romain François, Lionel Henry, and Kirill Müller. *dplyr: A grammar of data manipulation*, 2021b. URL https://CRAN.R-project.or g/package=dplyr. R package version 1.0.7.

Hadley Wickham et al. Elegant graphics for data analysis. *Media*, 35(211): 10–1007, 2009.

worldbank.org. *World Development Indicators*, 2021.

Kevin Wright. Revisiting Immer's barley data. *The American Statistician*, 67 (3):129–133, 2013.

CF Jeff Wu and Michael S Hamada. *Experiments: planning, analysis, and optimization*, volume 552. John Wiley & Sons, 2011.

Frank Yates. Complex experiments. *Supplement to the Journal of the Royal Statistical Society*, 2(2):181–247, 1935. ISSN 14666162. URL http://www. jstor.org/stable/2983638.

Index